# *The Language of Atoms*

Benedicto D.D.

*et Venus imminuit viris puerique parentum*
*blanditiis facile ingenium fregere superbum.*
*(De rerum natura 5.1017–18)*

# The Language of Atoms

## *Performativity and Politics in Lucretius'* De rerum natura

W. H. SHEARIN

OXFORD
UNIVERSITY PRESS

OXFORD
UNIVERSITY PRESS

Oxford University Press is a department of the University of Oxford.
It furthers the University's objective of excellence in research, scholarship,
and education by publishing worldwide.

Oxford New York
Auckland Cape Town Dar es Salaam Hong Kong Karachi
Kuala Lumpur Madrid Melbourne Mexico City Nairobi
New Delhi Shanghai Taipei Toronto

With offices in
Argentina Austria Brazil Chile Czech Republic France Greece
Guatemala Hungary Italy Japan Poland Portugal Singapore
South Korea Switzerland Thailand Turkey Ukraine Vietnam



Published in the United States of America by
Oxford University Press
198 Madison Avenue, New York, NY 10016



Library of Congress Cataloging-in-Publication Data
Shearin, W. H.
The language of atoms : performativity and politics in Lucretius' De rerum natura / W.H. Shearin.
pages cm
Includes index.
ISBN 978-0-19-020242-2 (hardcover : alk. paper) 1. Lucretius Carus, Titus—Language.
2. Lucretius Carus, Titus. De rerum natura. 3. Atoms. I. Title.
PA6495.S54 2014
187—dc23
2014010918

1 3 5 7 9 8 6 4 2
Printed in the United States of America
on acid-free paper

# *Contents*

# *Preface*

*Exi, <si> hodie nata, si ante nata, si hodie creata,*
*si ante creata;*
*hanc pestem pestilentiam, hunc dolorem,*
*hunc tumorem, hunc ruborem, has toles, has tosillas,*
*hunc panum, has panuclas, hanc strumam, hanc*
*strumellam*
*hac religione evoco, educo, excanto de istis membris*
*medullis.*

(MARCELLUS *Book on Drugs* 15.11)[1]

*Get out, whether born today or before, whether made*
*today or before;*
*this disease, pestilence, this pain,*
*this swelling, this redness, these goiters, these tonsils,*
*this abscess, these warts, this tumor, this inflammation,*
*with this ritual utterance (*hac religione*), I call, I lead,*
*I sing you out from these limbs and marrows.*

THIS ENCHANTMENT, PRESERVED by the fifth-century CE author Marcellus Empiricus, differs markedly from typical methods of modern Western medicine. It introduces us to an ancient, therapeutic speech act. "While fasting," Marcellus writes, "you will bewitch a fasting patient by holding the affected place with three digits—the thumb, middle finger, and ring finger—having raised up the other two." Then "you will speak" the lines above. This description provides us with only the barest details of a healing

1. For the text of Marcellus, I follow Niedermann-Liechtenhan 1968.

ritual, yet one thing seems clear: the enchantment aims to do something—to "call out" (*evoco*) and expel disease—and it aims to do it in the very act of being pronounced.

Lucretius' lengthy poem, *De rerum natura* ("On the Nature of Things"), does not represent as direct or obvious a therapeutic speech act as Marcellus'. In fact, if we limit speech acts solely to what John Austin called illocutionary acts, acts such as Marcellus' *evocatio* or the "I do" of a standard marriage rite in which an effect (marriage, healing) seemingly is accomplished by the very words of a ritual, it may not qualify as one at all.[2] Still, Lucretius' poem has often been suggested to have broadly therapeutic aims, whether or not it realizes these aims successfully or as part of a well-defined ritual context. *De rerum natura* is a poem with a pragmatic orientation, a poem that is centrally concerned with doing things to its readers. In this sense, we may usefully speak of Lucretius' work as performative, as not simply interested in describing the natural world but rather as centrally invested in doing things with those descriptions. The present study, which draws upon recent scholarship and direct readings of Lucretius' text, thus suggests that Lucretius' engagement with speech act theory is productive, both in that it opens up new ways of approaching and understanding *De rerum natura* and in that it allows insight into the creative, "productive" mechanisms of the poem. Speech acts, whether therapeutic or not, occur in conjunction with Lucretius' text in at least two different ways. On one level, we may speak of speech acts represented in *De rerum natura*—invocations, promises, and other acts, recorded directly or implicitly within Lucretius' text and serving a function within his "narrative" (that is, his presentation of Epicurean doctrine and the "story" of the natural world). In addition, we may also speak of another kind of speech act, of which Marcellus' therapeutic *evocatio* is an example. A speech act of this kind is not so much represented in the text as the effect of the text itself.

This work weaves back and forth between these two levels of speech acts, at times pursuing single, lexical items in Lucretius' text (e.g., *fides, foedus, terminus*) in an attempt to identify promises and acts of naming within *De rerum natura*. At other times, the study pursues the larger function of these speech acts within a cultural frame. Certain lexical items in Lucretius' text can reveal, when read aright, connections to contemporary political and societal

2. Edmunds 2001: 27–28, for example, distinguishes between speech acts and representations of speech acts, placing Lucretius only in the latter category. He is reacting against Carroll 1992: 106, who asserts that Lucretius' poem is an illocutionary act, as well as Beardsley 1970: 59–61. I am happy to allow that *De rerum natura* is not a proper illocutionary act, but the poem still does things.

concerns. Thus, from the pursuit of represented speech acts, the investigation turns to the political message of those representations as well as to the possibly therapeutic effect of that political message.

The first chapter of this study, which lays the theoretical foundation for later readings, dwells primarily upon fragments concerning Epicurus' thinking about logic and language. It argues that Epicurus had an explicit notion of performative language, a fact that motivates further investigation of performatives within the text of Lucretius' *De rerum natura*, our longest surviving Epicurean poem. The second and third chapters consider promising and naming, respectively. Each chapter suggests that there are certain nascent democratic elements—anonymity and material equality, to name two—in Lucretius' representation of these items. The final chapter then pursues these democratic elements with the result of describing a "catachrestic" politics within Lucretius' poem, a politics of instability and change, one that is at once more and less hopeful than a pure democracy.

However successful this final description, it cannot—nor does it claim to—determine the political effect of Lucretius' poem in any absolute way. Such effects may, in any case, have varied from reader to reader.[3] Yet the lens of performative language allows us to lay bare a few of the mechanisms through which Lucretius verbally creates and re-creates the natural and human worlds. Even if we do not always know precisely what they do, these mechanisms orchestrate political effects, and they therefore merit more than passing attention.

3. Compare the remarks of Kennedy 2013b: 51: "... the question of the politics of a particular text cannot be reduced to issues of authorial intention, but is caught up in issues of reception and appropriation—including our own—as it is accommodated to differing ideological and interpretative protocols, and in circumstances its author could not have foreseen...." See, too, Kennedy 1992 on "Augustan" and "anti-Augustan" interpretations.

# *Acknowledgments*

SCHOLARLY WORK, EVEN when conducted in relative isolation, is necessarily collaborative. Every book draws upon many others; every idea owes much to the contexts and circumstances that produce it. This work is no exception: many friends, family, and fellow scholars helped me, both directly and indirectly, as I struggled first to write and then to revise and refine it. It is a pleasure to acknowledge a few of these kindnesses.

This book began life as a Ph.D. dissertation, which was submitted to the University of California, Berkeley, in December 2007. Although it has been rewritten, revised, and updated in significant ways, the present version retains the basic structure and central arguments of the original dissertation. The committee that oversaw the earlier work—A. A. Long, Ellen S. Oliensis, Mark Griffith, and Kent Puckett—is due my sincerest thanks for important challenges and encouragement while the project was being formulated. All of them are versatile, interdisciplinary scholars whose teaching and writing offer formidable models of what academic work may be. Without their help and guidance, this book could never have gotten off the ground.

Subsequently, the Classics Department and the Andrew W. Mellon Fellowship of Scholars (an interdisciplinary humanities postdoctoral program) at Stanford University provided an ideal home for me to begin turning the dissertation into a book. Seth Lerer, the original director of the Mellon program, as well as Lanier Anderson and J. P. Daughton, its current directors, were all significant mentors, who helped me find my way as I was learning how to work and write while simultaneously navigating the expectations and responsibilities of a full-time faculty member. The other Mellon fellows—Heather Blair, Alex Cook, Kate Elswit, Brendan Fay, Shana Goldin-Perschbacher, Andrew Goldstone, Danielle Heard, Alan Mikhail, Bradley Naranch, Sarah Pourciau, Edith Sheffer, Joel Velasco, and Karen Zumhagen-Yekplé—provided essential encouragement and collegial support during my two years on the Farm. Within the Stanford Classics department, Maud Gleason, Christian Kaesser, and Richard P. Martin in particular offered important help and scholarly advice during my time in Building 110. In a more general sense, I also owe a significant, long-standing debt to the entire Stanford Classics faculty: without their guidance during my undergraduate years, I never would have pursued graduate education or entered the field of Classics professionally. Marsh McCall, Andrea Nightingale, and Susan Stephens especially have been generous with their time, help, and advice over a lengthy period.

The book reached its final form during my first few years as a faculty member at the University of Miami. My colleagues in Classics over that period—Scott Farrington, Jennifer Ferriss-Hill, George Hendren, John T. Kirby, Valentina Popescu, John Paul Russo, and Han Tran—all offered significant moral and intellectual support. John Paul Russo, in particular, read and commented upon earlier versions of the second chapter. Outside of Classics, Frank Palmeri, Maria Galli Stampino, Stephen Halsey, and Mihoko Suzuki have all been important interlocutors, teaching me through advice and example how to navigate the university. The vibrant Humanities Center at the University of Miami provided a collegial and stimulating intellectual environment in which to make final revisions to this manuscript.

Audiences in Stanford and Bristol heard earlier versions of material in the first chapter. I am grateful to both groups for enduring the preliminary state of the material on those occasions and for offering helpful commentary. In Stanford, Phil Horky, Bradley Naranch, Joel Velasco, and Mike Wigodsky all provided significant feedback and support. In Bristol, Kurt Lampe, Ika Willis, and Duncan Kennedy asked important questions and prodded me to think harder. Later, when the book was in full manuscript form, Duncan and another reader for Oxford University Press provided astute observations and recommendations for improvement. When the book was in its final stages, Barney Taylor provided insightful comments and challenges on the first chapter. I have not been able to pursue all of his suggestions, but wrestling with his thoughtful interventions made it possible to revisit the material with new eyes.

A number of people outside the formal walls of the academy also have worked to bring this book to life. At Oxford University Press, Stefan Vranka and Sarah Pirovitz have been efficient, helpful, and supportive at every turn. Bérénice Constans graciously allowed me to use her wonderful "Une boule de velours bien ronde qui n'aura jamais de lèvres" as my cover image. My copy editor, Patterson Lamb, did an excellent job with a messy, difficult manuscript, saving me from numerous bibliographical omissions and other errors. Eswari Marudhu and the production team at Newgen Knowledge Works have been detailed and attentive, ushering the manuscript through its various intermediate stages in a timely fashion.

On a more personal level, this book could not have been written without the help of my family. At many turns, my mother, Anne, provided childcare that enabled me to write. My father, Norman, and sisters, Faith and Dana, have listened with empathy and advised. My son's mother, Han Tran, has talked and worked with me over many years to ensure that I have been able bring this project to its end. For that and many other things, I owe her great thanks. Finally, my son, Benoît, has grown up alongside this work. It would be wrong to suggest that he has hastened its completion, but he has influenced it in ways he little knows; and he undoubtedly made the process of writing much more worthwhile. I dedicate the completed book to him.

# *Abbreviations*

| | |
|---|---|
| BNP | M. Beard, J. North, and S. Price. 1998. *Religions of Rome*. 2 vols. Cambridge. |
| CIG | 1828–77. *Corpus Inscriptionum Graecarum*. Berlin. |
| CIL | 1863–. *Corpus Inscriptionum Latinarum*. Berlin. |
| DRN | W. H. D. Rouse (ed. and trans.) 1992. *Lucretius: De Rerum Natura*. Cambridge, Mass. (Revised by M. F. Smith.) |
| DRP | Cicero *De re publica*, cited from J. G. F. Powell (ed.) 2006. *M. Tulli Ciceronis De re publica, De legibus, Cato Maior De senectute, Laelius De amicitia.* |
| LS | A. A. Long and D. N. Sedley (eds.) 1987. *The Hellenistic Philosophers*. 2 vols. Cambridge. |
| LSJ[9] | H. G. Liddell and R. Scott (eds.) 1968. *A Greek-English Lexicon*. Ninth Edition. Oxford. (Revised by H. Stuart Jones and R. McKenzie.) |
| OLD[2] | P. G. W. Glare (ed.) 2012. *Oxford Latin Dictionary*. Second Edition. Oxford. |
| RS | Epicurus *Ratae Sententiae (Master Sayings, Kuriai Doxai)* cited from T. Dorandi (ed.) 2013. *Diogenes Laertius: Lives of Eminent Philosophers*. Cambridge. |
| SSR | G. Giannantoni (ed.) 1990. *Socratis et Socraticorum Reliquiae*. 4 vols. Naples. |
| TLL | 1900–. *Thesaurus Linguae Latinae*. Leipzig. |
| Usener | H. Usener. 1887. *Epicurea*. Leipzig. (Reprint with Italian Translation and Notes by Ilaria Ramelli: Milan, 2002) |
| VS | Epicurus *Vaticanae Sententiae (Vatican Sayings)*, cited from T. Dorandi (ed.) 2013. *Diogenes Laertius: Lives of Eminent Philosophers*. Cambridge. |

# I

# *Speech Acts in Epicureanism*

*... the principal burden of his [sc. Catullus'] contemporaries' compositions—Lucretius, Cicero, Varro Atacinus—tends to be descriptive, that is to record res gestae or explain the phenomena of nature ...*[1]

*The act, an enigmatic and problematic production of the speaking body, destroys from its inception the metaphysical dichotomy between the domain of the "mental" and the domain of the "physical," breaks down the opposition between body and spirit, between matter and language.*[2]

## *I. Locating Speech Acts in Lucretian Scholarship*

At least since the nineteenth century, the critical reception of Lucretius has been marked by a double gesture. On the one hand, the Epicurean poet, perhaps as much as any classical author, has been the subject of intense textual scrutiny. Most famously, with his monumental edition and commentary of 1850, Karl Lachmann developed his method of stemmatic criticism against

1. Selden 1992: 481–82, now reprinted (in an expanded and updated form) at Selden 2007: 522–23. Selden adds in an accompanying note: "Note the titles of the major poetic projects of Catullus' generation: *De rerum natura, De consulatu suo, De temporibus suis, Marius, Pragmatia Belli Gallici, Chorographia, Bellum Sequanicum* ..." (1992: 503n100; 2007: 523n107). By looking at the title of Lucretius' poem, he suggests, we may recognize that his project, unlike that of Catullus, was chiefly descriptive, not performative. This book contests such a view.

2. Felman 2002: 65. Emphasis original.

the text of Lucretius.[3] Lachmann dazzled his contemporaries with his physical reconstruction of the archetype, or ancient copy from which all extant manuscripts of Lucretius descend. Yet he was hardly the only scholar of his era interested in careful study of the words of *De rerum natura*.[4] Writing in the 1820s and 1830s, Johann Caspar von Orelli and Johan Madvig already had anticipated many of Lachmann's results through careful study of the witnesses to Lucretius' text.[5] Jacob Bernays, spurred on by a prize offered at the University of Bonn in 1845, and Hugo Purmann, writing almost simultaneously, also addressed the criteria for establishing a properly critical edition of the Epicurean poet.[6] Such textual scrutiny has continued throughout the twentieth century and into the twenty-first: Richard Heinze, Hermann Diels, Alfred Ernout, Cyril Bailey, Edward J. Kenney, Konrad Müller, Michael Reeve, and David Butterfield constitute only some of the most prominent names within this ongoing tradition of carefully adjudicating the manuscripts and establishing the text of Lucretius.[7]

Alongside this active tradition of textual scholarship, another form of reading has arisen, one that—even if it takes its cues from *De rerum natura* itself—focuses instead upon the author and the "outside" of his text, the supposed historical circumstances surrounding the composition of his poem. From Jerome to Cyril Bailey, biographical speculation not only has been present in the reception of Lucretius, but it has occupied a central position in that reception.[8] The fourth edition of the *Encyclopaedia Britannica*, published in 1810, reports that Lucretius "fell into a frenzy, occasioned by a

3. Lachmann 1850: 1–10 discusses Lucretius' archetype. Timpanaro 2005 (dating back to work from the 1950s) remains the best discussion of Lachmann's place in the development of stemmatic textual criticism.

4. Passannante 2011: 78–119 offers a stimulating meditation on the interplay of (Epicurean) materialist thought with the material reconstruction of Lucretius' poem.

5. Orelli 1827, Madvig 1832. Timpanaro 2005: 102–18 remains the best discussion of the interrelationship of Lachmann, Orelli, Madvig, Bernays, and Purmann.

6. Bernays 1847, Purmann 1846.

7. See, e.g., Heinze (ed.) 1897, Diels (ed.) 1923, Ernout (ed.) 1978, Kenney (ed.) 1984, Müller (ed.) 1975, Reeve 1980, Reeve 2005, and Butterfield 2013. Similarly textual is scholarship influenced by the New Criticism (e.g., West 1969, several of the chapters in Dudley (ed.) 1965) as well as writing on soundplay and wordplay in *De rerum natura* (e.g., Friedländer 1941).

8. For Jerome's famous statement, see *Chronica* s. a. Abr. 1923 (94 BCE): *T. Lucretius poeta nascitur. qui postea amatorio poculo furorem versus cum aliquot libros per intervalla insaniae conscripsisset, quos postea Cicero emendavit, propria se manu interfecit anno aetatis XLIIII.* ("Titus Lucretius, the poet, is born. Later when he, rendered mad by a love potion, had composed several books—which Cicero afterwards corrected—through fits of insanity, he died by his own

philtre given him by his wife, who was distractedly fond of him"; that "during the intervals of his madness, [he] put Epicurus' doctrines into verse"; and "that he killed himself in a fit of madness ... when 51 years old."[9] Examples may be multiplied. Although he casts doubt upon the love potion itself, Johannes Mewaldt writes in his 1927 Pauly-Wissowa article that "occasional madness" (*zeitweilige geistige Umnachtung*) and "suicide at a relatively young age" (*Selbstmord in verhältnismäßig frühem Alter*) may "count as facts" (*als Tatsachen gelten*).[10] Writing only a few years later, Otto Regenbogen more skeptically proclaims the biography of Jerome "more or less entirely uncertain" (*so gut wie alles unsicher*).[11] Yet he preserves the portrait of Lucretius as a melancholic loner.[12]

More recently, scholars have dismissed purely biographical speculation, but they have not given up on articulating the relationship of Lucretius' text with his historical "outside."[13] Benjamin Farrington's *Science and Politics in the Ancient World*, J. H. Nichols, Jr.'s *Epicurean Political Philosophy*, and J. D. Minyard's *Lucretius and the Late Republic*, for example, all treat—in different ways—the political significance of Roman Epicureanism.[14] Such works made it possible for Don Fowler to comment over twenty years ago that "[i]t is becoming a commonplace of modern scholarship that the *De Rerum Natura* is a political work."[15] If, then, the ghost of Lucretius himself has largely been

---

hand in his forty-fourth year.") For Bailey's, see Bailey (ed.) 1947: 12: "There is nothing in the poem which makes the idea of 'morbid depression' impossible, a good deal which supports it. The *insania* and the suicide are not inconsistent with anything that we can gather of the poet's character, and the story of the love-philtre, though in no way supported, may also be true."

9. Passages excerpted from the fuller citation at Johnson 2000: 103.

10. Mewaldt 1927: 1659.

11. Regenbogen 1932: 20.

12. Regenbogen 1932: 15: "... Lucrez ist einsam.... Lucrez ist allein, mit seiner Sehnsucht aus dem Dunkel in das Helle strebend, glühend von seinem Erleben, dürstend, es werbend mitzuteilen, leidenschaftlich ringend um den Freund und Gönner, überströmend dankbar gegen den Erlöser."

13. A fuller discussion of the older scholarly tradition on Lucretius' biography may be found at Bailey (ed.) 1947: 8–12. More recent studies include Canfora 1993 (which examines historical evidence from the Republican and early Augustan periods) and Holford-Strevens 2002. The Teubner edition of Martin (Martin 1963) usefully collects the *testimonia vitae*, including literary references to Lucretius in antiquity.

14. Farrington 1965 (originally published 1939, reviewed critically by Momigliano 1941, angrily by Guthrie 1940), Nichols 1976, and Minyard 1985. Other political treatments of Lucretius include Fallot 1977 and Grimal 1978. The best recent works connecting Lucretius to politics are Schiesaro 2007a, 2007b and Kennedy 2013b.

15. Fowler 1989: 122.

exorcised from the annals of scholarly writing—being summoned occasionally only to reinforce that very exorcism—he has been replaced by the ghosts of his contemporaries, the ghosts he aimed to save with his Epicurean message.[16] "The poem addresses," Fowler writes, "the ills of contemporary society and suggests a solution for them in Epicurean philosophy."[17]

As Fowler indicates, recent scholarship has connected *De rerum natura* to its "outside" by emphasizing the therapeutic potential of the poem. Nor is such interpretation confined to political readings of Lucretius. Ancient philosophy has become increasingly hospitable to reading philosophical texts as part and parcel of a way of life, a life that demands psychic therapy or *exercises spirituels*.[18] Perhaps the most elaborate English-language characterization of Hellenistic philosophy (and Lucretius specifically) as therapeutic is Martha Nussbaum's *The Therapy of Desire*.[19] That work lays out *in extenso* the import of the medical model for philosophy, commenting that on such a model "[a]rguments have a *practical goal*: they are directed at making the pupil better."[20] Unlike earlier biographical speculation, this strain in scholarship, which chiefly aims at explaining philosophical arguments, tends to leave Lucretius' contemporaries largely as shadows, pointing rather to the constant political strife of the later Roman Republic, which Lucretius aimed to ameliorate.[21] Such work has been colored more by a simple gazing "outside" of

---

16. See Holford-Strevens 2002, which states the now dominant skeptical position on the Lucretian biographical tradition, drawing in particular upon Ziegler 1936.

17. Fowler 1989: 122.

18. See, e.g., Foucault 1984, Hadot 1987, and Cooper 2012. One could add further references.

19. In French, Voelke 1993 anticipates some of Nussbaum's claims. Still earlier are Gigante 1975 (on the Epicurean Philodemus) and Kassel 1958: 29–32 (on Epicurean *consolatio*). Pigeaud 1981 studies psychic therapy in ancient medicine as well as ancient philosophy. The medical connection is made explicit in Lucretius' poem. The famous honeyed-cup simile found at *DRN* 1.926–50 and 4.1–25, for example, conjures up the notion of the poem as performing a therapeutic function. Epicurus, too, supports a therapeutic understanding of philosophy: κενὸς ἐκείνου φιλοσόφου λόγος, ὑφ' οὗ μηδὲν πάθος ἀνθρώπου θεραπεύεται [Porphyry *Ad Marc.* 31, p. 209, 23 Nauck = Usener 221, part (Usener 1887: 169)] ("Empty is the word of that philosopher by whom no human suffering is alleviated . . .").

20. Nussbaum 1994: 46.

21. A partial exception: there has always been biographical speculation, not entirely without evidentiary support, about Gaius Memmius, the addressee of Lucretius' poem (e.g., Münzer 1931 and, for a less strictly biographical discussion of Memmius as "an example of the worst in a bad society," Roller 1970). Yet recent work has turned away from trying to reconstruct Memmius' biography and instead endeavored to use him as a foil, whether negative or positive, for the reader, whether implied or actual. See Clay 1983a: 214–15 and *passim*, Mitsis 1993, Fowler 2000, and Gale 2004.

the text (as a means to justify argumentative orientation) than by detailed, specific examination of political, historical, or social context.[22]

Increasingly, then, scholarship on Lucretius has moved from describing the historical circumstances surrounding the composition and transmission of *De rerum natura* toward characterizing the ways in which the poem acts (or has acted) within history.[23] Detailed textual scrutiny and historical reconstruction have joined forces to examine textual effects. The "inside" of the text, so to speak, has gotten "outside"; and scholarship asks ever more how this happens. Within this larger conversation, it is odd that speech act theory and its later appropriations have received only limited attention, precisely because they address themselves to the linguistic mechanisms of transactions between inside and outside. In fact, speech act theory was conceived to solve a specific version of the inside-outside problem. Its founder, J. L. Austin, in his now canonical *How to Do Things with Words* (first delivered as the William James Lectures at Harvard in 1955), explores precisely how using language can be a form of doing, that is, how language can extend beyond functioning as a descriptive tool and instead act "outside" of language.[24] The opening of that work articulates the problem as follows:

> It was for too long the assumption of philosophers that the business of a "statement" can only be to "describe" some state of affairs, or to "state some fact", which it must do either truly or falsely.... But now in recent years, many things which would once have been accepted without question as "statements" by both philosophers and grammarians have been scrutinized with new care.... It has come to be commonly held that many utterances which look like statements are either not

22. This brief survey is hardly exhaustive. It highlights strands of scholarship that are both salient and important for the present study. For more thorough studies of Lucretian bibliography, see Gordon 1962, Kenney 1977, the introduction to Classen (ed.) 1986, Erler 1994, and the introduction to Gale (ed.) 2007. Gillespie and Hardie (eds.) 2007 leans heavily towards reception, providing a useful thirty-page (327–56) collected bibliography.

23. The above discussion focuses on readings of Lucretius as a therapeutic poet, but other reading strategies may be seen as connecting the poet to his "outside." Focusing on the (persuasive) rhetorical strategies of the text (e.g., Asmis 1983, Markovic 2008), viewing portions of it as Epicurean tests (Clay 1983a), or defining the "reader-addressee" (Conte 1994a: 18–34) as a model for all readers raises issues of pragmatic force, of how the text acts upon its "outside."

24. In the wake of Austin, we may question whether language ever could be said to function purely as a descriptive tool, but it is convenient to present the development of his theory in these terms. For an insightful discussion of the ways in which the Roman teacher of rhetoric Quintilian already recognized the questionable status of purely descriptive ("constative") discourse, see Arweiler 2010 (especially 204–11).

> intended at all, or only intended in part, to record or impart straightforward information about the facts . . .[25]

To discuss this observation that all statements are not merely descriptive, Austin introduces a new vocabulary. On the one hand, he calls "constatives" utterances understood in the traditional manner, that is, as descriptions that may be characterized as true (correct) or false (incorrect). On the other, he terms "performatives" utterances that—rather than describing a state of affairs—do something in their very act of being uttered.[26] He offers a few examples of the second category:

(E. *a*) "I do (*sc.* take this woman to be my lawful wedded wife)"—as uttered in the course of the marriage ceremony.

(E. *b*) "I name this ship the *Queen Elizabeth*"—as uttered when smashing the bottle against the stern.

(E. *c*) "I give and bequeath my watch to my brother"—as occurring in a will.

(E. *d*) "I bet you sixpence it will rain tomorrow."[27]

These examples, each of which may be paralleled in Latin (e.g., Latin *sponsio* or *stipulatio*), show that often the most banal utterances are not mere statements of fact.[28] Whatever such utterances may mean, their chief significance lies in what they do (creating a marriage, giving a legal name, etc.).[29] Austin's insight, although perhaps nothing more than the articulation of an idea that many societies have long known implicitly, nonetheless has potentially

25. Austin 1975: 1–2.

26. The vocabulary used here ("performative" vs. "constative") is the initial one that Austin develops in *How to do Things with Words*. It has had the greatest influence both in subsequent discussions of Austin's work and in developments of that work. Austin also develops further terminology to aid in refining and rendering more specific the performative-constative distinction. Among other classifications, he introduces the division locutionary-illocutionary-perlocutionary as well as the more elaborate (and "rebarbative") classificatory system of his final lecture (Lecture 12 at Austin 1975: 148–64).

27. Austin 1975: 5.

28. On *sponsio* (or *pactio*), the Roman engagement, see Treggiari 1991: 170–80. On *stipulatio*, the Roman verbal contract, see Kaser 1971: 538–43 (§128). The Roman wedding ritual now has a detailed study: Hersch 2010.

29. One common way of articulating the connection between meaning and doing in performative utterances is to say, with Émile Benveniste, that performatives are self-referential (*sui-référentiel*), that is, that they constitute that to which they refer. See Benveniste 1966: 274.

far-reaching implications for the study of a poet such as Lucretius, whose work—at least on the current scholarly understanding—is conceived of as acting therapeutically. Speech act theory offers a lens through which to scrutinize the pragmatic force of a text whose pragmatic orientation is remarked so frequently.[30]

Placing Lucretius into conversation with Austin thus appears to hold promise for addressing long-standing scholarly concerns. In its subsequent chapters, the present work attempts to foster such conversation. But further preliminary specifications and qualifications are needed. Since its inception with Austin, speech act theory has branched out into a vast array of related, if incredibly diverse, fields.[31] To name only a few of the most well-known instances, the work of philosophers such as John Searle, Stanley Cavell, and Jacques Derrida as well as that of gender-theorist Judith Butler and literary critics such as Shoshana Felman, Paul de Man, and J. Hillis Miller have all benefited from, and directly engaged with, the central realization of speech act theory, namely, that language may be performative as well as constative. It would thus require an entire book—if not several books—to catalogue and examine the various ways in which speech act theory has been received.[32] Yet any present-day study that takes advantage of the insights of speech act theory cannot, and should not, ignore these more recent developments. As something of a compromise, this introduction points to a single important issue—what one, following earlier discussions of speech act theory, may call "context"—which lays the groundwork for later discussion, without

---

30. In addition to the poem's therapeutic orientation, Lowrie 2009: 33 remarks that "Lucretius' didacticism lends his poem a pragmatic aim."

31. Loxley 2007 provides a convenient overview of the various directions in which speech act theory has developed.

32. The seminal tract for speech act theory (as noted above) is Austin 1975 (which derives from work Austin had initiated at least as early as 1939—see, e.g., the early paper "The Meaning of a Word" in Austin 1979). The vast literature on speech act theory includes Searle 1969, Derrida 1988, de Man 1979, Felman 2002, Cavell 1995, and Miller 2001 and 2005. A version of speech act theory, mostly filtered through Derrida and gender scholar Eve Kosofsky Sedgwick (see Parker and Sedgwick (eds.) 1995: 1–18), also informs Judith Butler's work. See, for example, Butler 1993: 12–16, especially nn9–10, where she lists some of her points of reference, and her "Afterword" to Felman 2002. Within the study of Latin poetry, important discussions of speech act theory include Selden 1992 (which draws heavily upon de Man 1979) and Lowrie 2009 and 2010. Readers are also advised to consult the massive online bibliography devoted to "Discussions of John Austin and His Writings," hosted by UC-Irvine: http://www.lib.uci.edu/about/publications/philosophy/austin/works_about.html (accessed 14 March 2013). Culler 2000 (republished in Culler 2007) and Miller 2007 have begun sorting out the various uses of the performative in academic discourse.

attempting to provide a comprehensive account of all that is, or could today be called, speech act theory.

In particular, Derrida's account of context within speech act theory offers important resources for staging the encounter between speech act theory and Lucretius' poem. The French philosopher revises two positions that are often, rightly or wrongly, attributed to Austin. On the one hand, he argues forcefully that speech act theory should not be based exclusively upon spoken discourse. While Austin's initial examples of performativity, as we have seen, include written items such as wills, speech acts in their earliest incarnations are associated strongly with the first of Austin's examples, the marriage rite, which is performed by live, embodied voices in a formal, ritual context.[33] In his discussion of context, Derrida aims to give an account of how language (or locution) in general—including written, literary language—functions and acts. On the other hand, Derrida also questions the purity of the performative-constative distinction, suggesting that one must think of language as both acting and describing, often simultaneously. While he in no way discards performativity as a concept, he questions its isolation from other functions of language. For a written text like *De rerum natura*, which does not obviously derive from a ritual context and which at times seems to expound the nature of the universe descriptively, Derrida's interventions help one understand how speech act theory may come into play.

Derrida brings together these two points—that speech act theory may apply to writing and that performativity and constativity are not mutually exclusive categories—in the key larger argument of his study of context. Context is not, as it sometimes may seem, a fixed entity that singularly determines linguistic meaning. Nor is it only in play in spoken, ritual discourse. Instead, context is at once static and dynamic, serving as both the firm ground upon which meaning is based and the ever-changing frame that allows linguistic and social evolution. Such a statement is not tantamount, as critics of Derrida are wont to claim, to the assertion that language, which depends upon context for meaning, is meaningless.[34] Rather, any full description of

---

33. Eve Kosofsky Sedgwick (1993: 3), in pointing out the centrality of the marriage rite to Austin's work, remarks: "The marriage ceremony is, indeed, so central to the origins of 'performativity' ... that a more accurate name for *How to Do Things with Words* might have been *How to say (or write) 'I do' about twenty million times without winding up any more married than you started out.* (Short title: *I Do—Not!*)."

34. Petrey 1990: 152, who provides an insightful juxtaposition of Austin and Derrida, makes clear how language remains meaningfully rooted in social convention even after Derrida's critiques: "Deconstructionists' refutations of Austin's opposition between literary and serious language, like their demonstration that constative and performative come together in locution

language, he suggests, must take account of context and its simultaneous constancy and dynamism. And Derrida is not alone in making this point. In important ways, as we shall see, Epicurus, or at least Epicurean-influenced writing, was there before him.

## *II. Context in Speech Act Theory: Austin, Derrida, and their Critics*

That context figures importantly in any discussion of speech act theory has been recognized from Austin onward. Indeed, it is the very centrality of context in the study of speech acts that makes such study a way of considering the inside-outside problem. By locating a theory of language within a theory of action, speech act theory considers how what seems interior and personal, the production of language, acts outside in the context of the larger world. Austin concentrates his own discussion of context primarily within a consideration of felicity, a consideration, that is, of whether a particular speech act succeeds in acting in a specified way. In the third lecture of *How to Do Things with Words*, for example, where he discusses in a general fashion the necessary conditions for a felicitous speech act, he lays down as one of his rules: "[t]he particular persons and circumstances in a given case must be appropriate for the invocation of the particular procedure invoked."[35] He adds a list of examples in which the infelicity of a speech act (that is, its "unhappiness" or "failure") arises through the violation of this rule: "'I appoint you,' said when you have already been appointed, or when someone else had been appointed, or when I am not entitled to appoint, or when you are a horse: ... 'I give,' said when it is not mine to give or when it is a pound of my living and non-detached flesh."

Austin's ironic lists make clear that it matters who performs a speech act and how it is performed. Context, in other words, is essential to the success of speech acts. Yet his discussions do not consider, perhaps intentionally, the

---

as well as illocution, complicate speech-act theory *without at all invalidating it.* There are more reasons than Austin knew for attacking the descriptive fallacy and for bringing saying and doing together. But those reasons leave intact the vision on which speech-act theory is grounded, *recognition that language performs what its users live by virtue of its articulation with the conventions they observe*." (Emphasis added.)

35. Austin 1975: 34. This criterion (labeled "A. 2") gives a good sense of the role of context for Austin, but the full list of felicity conditions discussed in Austin's third lecture (1975: 25–38) may be subsumed under the broader heading of "context."

question of context in its most general form, a topic that has been addressed more fully in the work of Jacques Derrida, particularly in his debates with John Searle.[36] These debates stem from an essay by Derrida «signature événement contexte» ("Signature Event Context"), originally composed for delivery at the Congrès international des sociétés de philosophie de langue française ("International congress of societies of French-language philosophy") in August 1971, which was translated into English and published in the journal *Glyph* in 1977. In the same issue of *Glyph*, John Searle issued a reply ("Reiterating the Differences: A Reply to Derrida") to the English version of Derrida's essay, in which he tries to indicate both ways that Derrida failed to understand Austin and, apart from any consideration of Austin, failings in Derrida's arguments. Derrida later wrote a reply to this reply, entitled «Limited Inc a b c … » ("Limited Inc abc …"), which was translated into English and first published in *Glyph* 2 (1977) and then re-published together with "Signature Event Context" in a volume called *Limited Inc* (1988). This at times hostile exchange is relevant to the present discussion precisely because, as the title of Derrida's first piece indicates, it takes context in speech act theory as one of its primary areas of focus.

Rehearsing all the arguments, verbiage, and attacks on both sides of the Searle-Derrida debate would, if instructive, involve us in a project well beyond that of the present book.[37] Focusing instead upon Derrida's original, explicit claims about context in speech act theory in "Signature Event Context" as

---

36. Austin, however, does allude to the need for a "general theory" of speech acts (Austin 1975: 148). He is a notoriously slippery and ironic writer, and it is in keeping with the tone of his extant philosophical work (which often takes as its target, on the one hand, what he sees as vacuous metaphysical generalization and, on the other, the logical positivism of philosophers such as A. J. Ayer [cf. Austin 1962a]) that he would not easily generalize about the nature of the context in speech acts. Such an observation may not exculpate him entirely from the charges that Derrida levels against him; however, any critique of Austin through the lens of Derrida must proceed with the recognition that Austin's project is, by its very nature, closely allied with Derrida's. Austin himself remarks at one point (Austin 1979: 182) that his way of doing philosophy might be called (in a manner less misleading than "ordinary language" or "analytic") "linguistic phenomenology," which is of course precisely what Derrida, beginning with his early critique of Husserl, practices. Shoshana Felman (in Felman 2002) is perhaps the reader who makes Austin look most like Derrida.

37. The summary presented here only considers the portion of the Searle-Derrida debate immediately relevant to speech acts (and specifically with "context" in speech acts). Searle subsequently used a review (Searle 1983) of Culler 1983 to launch a more general attack upon Derrida and deconstructive philosophy. There he famously recounts the following anecdote: "Michel Foucault once characterized Derrida's prose style to me as '*obscurantisme terroriste*.' The text is written so obscurely that you can't figure out exactly what the thesis is (hence '*obscurantisme*') and then when one criticizes it, the author says, '*Vous m'avez mal compris; vous êtes idiot*' (hence '*terroriste*')." Such comments give a taste of the level of acrimony the debate reached.

well as upon critiques of those claims is more manageable if also more dangerous, insofar as one runs a greater risk of mischaracterizing the views of both Derrida and his critics within a restricted treatment. But, in a certain sense, because it abuts many of the other considerations within Derrida's essay, consideration of the notion of context provides a substantial, if limited, glimpse of the whole: indeed, Derrida proclaims quite early in "Signature Event Context" that the essay "will be concerned with the problem of context and with the question of determining exactly how writing relates to context in general."[38] So, at least from Derrida's perspective, a focus upon context is not unwarranted. Let us turn, then, to Derrida and "Signature Event Context," the essay that sparked intense subsequent debate.

Although "Signature Event Context" begins with an epigram from Austin and inspires Searle's response through its consideration of the well-known ordinary language philosopher, the opening discussion focuses not on Austin but, in a manner somewhat reminiscent of Austin's own philological excursions, on a consideration of the meaning or meanings of the word communication (*communication*).[39] Yet at the same time that Derrida considers what communication means, he considers what it means to mean, that is, whether or not "communication" communicates "a determinate content, an identifiable meaning, or a describable value."[40] He asserts that—even to query the meaning of "communication"—he has been forced "to anticipate the meaning of the word *communication*," "to predetermine communication as a vehicle, a means of transport or transitional medium of a *meaning*, and moreover of a *unified* meaning."[41] Consideration of this question leads Derrida to the conclusion: "it seems self-evident (*aller de soi*) that the ambiguous field of the word 'communication' can be reduced massively by the limits of what we call a *context*."[42]

To this point, Derrida reaches a familiar philological conclusion, one represented on every page of works such as *A Greek-English Lexicon* of Liddell,

38. Derrida 1988: 2. (Translated by S. Weber and J. Mehlman.)

39. It may seem puzzling that a consideration of speech acts begins with a study of communication in general, but it should be noted that Derrida is, as will become clearer in the present discussion, an advocate of what is sometimes called an "undifferentiated textuality." In other words, he believes that any account of how language functions in, say, performative utterances must be able to take account of how the structure of language functions more generally.

40. Derrida 1988: 1. (Weber-Mehlman translation.)

41. Derrida 1988: 1.

42. Derrida 1988: 2. (Weber-Mehlman translation, slightly modified.)

Scott, and Jones and the *Oxford Latin Dictionary*: "context determines meaning." From here, however, he makes an unfamiliar turn: he proclaims "[b]ut are the conditions [*les réquisits*] of a context ever absolutely determinable? ... Is there a rigorous and scientific concept of *context*? Or does the notion of context not conceal ... philosophical presuppositions of a very determinate nature?"[43] With these leading questions, Derrida suggests that the "current concept of context" is insufficient. As he writes elsewhere,

> ... no meaning can be determined out of context, but no context permits saturation. What I am referring to here is not richness of substance, semantic fertility, but rather structure: the structure of the remnant of iteration.... I have given this structure many other names, and what matters here is the secondary aspect of nomination. Nomination is important, but it is constantly caught up in a process that it does not control.[44]

By "nomination" here, Derrida means the speech act of naming. Its secondary aspect is the fact that it not only gives a name, but that in the very act of naming it does something: it creates its own referent, thus changing its "context" (from the inside out, one might say) and preventing "saturation." These remarks may serve as a preface to Derrida's direct engagement with Austin, to which we now turn.

Through his analysis of Condillac and Husserl in "Signature Event Context" (which immediately precedes his discussion of Austin), Derrida presents a problem to which he often refers as "the classical conception of writing," that is, a notion of writing as providing a moment of "consciousness," a notion of writing as making fully present the intentions of the author. Derrida opposes this conception of writing for several reasons, not only because it, on his view, subjugates writing to speaking—that is, it renders writing nothing more than a transcript of the intention-driven speech the author would provide were he not absent—but also because it fails to take account of what Derrida often calls iterability, that is, the fact that writing "*must be able* to function in the absence of the sender, the receiver, the context of production, etc...."[45] Austin, too, despite "his patient, open, aporetic analysis" in *How to*

43. Derrida 1988: 2–3. (Weber-Mehlman translation.)

44. Derrida 1979: 81. (Translated by J. Hulbert.)

45. Derrida 1988: 48. (Translated by S. Weber.)

*Do Things with Words* and despite many virtues in his writing, succumbs—at least at times—to this classical conception of writing:

> Austin did not take account of what—in the structure of *locution* (thus before any illocutionary or perlocutionary determination)—already entails that system of predicates I call *graphematic in general* [*graphématiques en général*] and consequently blurs [*brouille*] all the oppositions which follow, oppositions whose pertinence, purity, and rigor Austin tried in vain to establish.[46]

The nature of language in general, Derrida suggests, blurs the central division at the heart of speech act theory. Austin cannot maintain absolutely or in all circumstances the performative-constative partition. Derrida continues:

> In order to show this, I shall consider as known and self-evident the fact that Austin's analyses require at all times a value of *context*, and even of a context able to be determined exhaustively, either in theory (*en droit*) or teleologically; and the long list of various "infelicities" which may affect the performative event always comes back to an element of what Austin calls the total context. One of those essential elements—and not one among others—remains, classically, consciousness, the conscious presence of the intention of the speaking subject in the totality of his speech act. As a result, performative communication becomes once more the communication of an intentional meaning, even if that meaning has no referent in the form of a thing or of a prior or exterior state of things.[47]

To buttress these assertions, Derrida selects a few passages from *How to Do Things with Words* in which Austin seems to commit himself to a classical view of writing. In particular, he fastens upon Austin's various exclusions in developing his notion of the felicity conditions of a given speech act, items that—for one reason or another—Austin leaves out of consideration. The

46. Derrida 1988: 14. (Weber-Mehlman translation, slightly modified.)

47. Derrida 1988: 14. (Weber-Mehlman translation, modified.)

most prominent exclusion—and the one to which Derrida objects most strenuously—runs as follows:

> ... as *utterances* our performances are *also* heir to certain other types of ill, which infect all utterances. And these likewise, though again they might be brought into a more general account, we are deliberately at present excluding. I mean, for example, the following: a performative utterance will, for example, be *in a peculiar way* hollow or void if said by an actor on the stage, or if introduced in a poem, or spoken in soliloquy. This applies in a similar manner to any and every utterance—a sea-change in special circumstances. Language in such circumstances is in special ways—intelligibly—used not seriously, but in many ways *parasitic* upon its normal use—ways which fall under the doctrine of the *etiolations* of language. All this we are *excluding* from consideration. Our performative utterances, felicitous or not, are to be understood as issued in ordinary circumstances.[48]

Derrida is bothered by the notion that certain forms of language should be termed, in an attempt to study language or speech acts generally, "non-serious" or "parasitic," precisely because such distinctions are based upon intention, which—as we have already noted above—Derrida denies can ever be fully present, in any act of speaking or writing. It is not that intention does not exist but rather that intention, however essential it may be to meaning, cannot solely determine it (and intention thus cannot adequately describe the way a linguistic system, in its totality, actually functions).[49] Terms such as "parasitic" and "non-serious" as well as the "ordinary" in "ordinary-language philosophy"—which all try to define and delimit context—are in Derrida's view never fully authorized:

> The concept of—or the search for—the context thus seems to suffer ... from the same theoretical and "interested" uncertainty as the concept

48. Austin 1975: 21–22.

49. Derrida's resistance to intention and intentionality stems from his reading of, and engagement with, psychoanalysis, in particular with the Freudian Unconscious. Within the Derrida-Searle debate, see Derrida 1988: 73–74: "Each time that the question of the 'ethical and teleological discourse of consciousness' arises, it is in an effort to uncover and to break the security-lock which, from *within the system* ... condemns the Unconscious as one bars access to a forbidden place." For further discussion, see Miller 2001: 94–98.

> of the "ordinary," from the same metaphysical origins: the ethical and teleological discourse of consciousness.[50]

Derrida's critique of Austin stems in part from what might be re-phrased as a subject-object problem. The so-called classical understanding of writing views writing as a tool for conveying unproblematically the subject's perspective, the unmediated intentions of the author. Derrida, on the other hand, is interested in specifying objectively the possibilities inherent in the medium of writing. He thus considers it necessary to understand and explain the possibilities of all written discourse, not just that which functions in the way the subject intends. Of course, "objectively" here is ripe for its own deconstruction: Derrida would hardly subscribe to the notion of a clear, simple opposition between subject and object. But, as long as we take "objectively" in a larger sense, namely, as encompassing already the larger system of subject and object, viewing Derrida's critique of Austin in these terms allows some sense of where their differences arise.

To some, Derrida's critique of Austin on the grounds of his use of "parasitic" seems unjustified. Searle, for example, suggests, among other things, that Austin's exclusion of parasitic forms is merely "a matter of research strategy."[51] Stanley Cavell insists that Austin's exclusion refers to his work on pretending: Austin excludes parasitic discourse here because he treats it elsewhere.[52] Richard Rorty approves Stanley Fish's interpretation that "*How to Do Things with Words* 'say[s] (about language, if not about philosophy) pretty much what Derrida himself wants to say.'"[53] On any of these views, there is little sense in Derrida's critique of Austin. It may be so. But if there is a difference between Austin and Derrida, at least as regards the notion of context in speech act theory, it lies in Derrida's explicit formulation that speech acts change the context in which they enter, as Derrida's defenders will point out.[54] On this issue, J. Hillis Miller has commented:

> ... Derrida wants to have it both ways. The context is there already, but it becomes a context only when the speech act intervenes within

50. Derrida 1988: 18. (Weber-Mehlman translation.)

51. Searle 1977: 205.

52. Cavell 1995: 57–60. For Austin's work on pretending, see Austin 1979: 253–71.

53. Rorty 1991: 86n3 referring to Fish 1982.

54. It may be possible to extract this position from Austin's text, yet the philosopher most often presents himself as stumbling upon problems in the formulation and understanding of "context" rather than acknowledging that speaking may act upon and change context.

> it, however weakly and without power to saturate it. The speech act nevertheless transforms the context it enters, even though in retrospect that context seems to have been there already as the ground of the speech act's efficacy.[55]

It is this notion of context, a context that is not fixed but at once determining of and determined by the speech act (and all the while never saturated), which is relevant to the discussion of speech acts in Epicureanism. Whether this idea is the brainchild of Derrida alone or of both Derrida and Austin, it is fortunately not our burden to decide.

## *III. Epicurean Speech Acts: Epicurus' Thinking on Language and Logic*

### (i) Situating the Question

If context, as considered above, is central to speech act theory, what does this claim mean for Lucretius? Earlier discussion has highlighted the fact that the work of the Epicurean poet is read regularly as therapeutic. Yet it is not only this practical goal that links *De rerum natura* to performative language and its contextual conditions. Indeed, the philosophical links between Lucretius and speech act theory run, I suggest, quite deep. While investigations of speech act theory often proceed from the supposition that the theory is universally valid, in the case of Epicureanism there is reason to believe that claims of speech act theory may be native to the philosophical school.[56] Of course, Epicurus and his disciples can hardly be accused of developing in every detail a theory that was not well articulated until the second half of the twentieth century, but Epicurean thinking on language anticipates, I suggest, the primary contribution of speech act theory to modern studies of language and logic—the recognition and articulation of the performative. It also may grasp nuances about context in speech act theory.

---

55. Miller 2001: 111.

56. Recent writing on Epicurean linguistic thought adumbrates this reading of Epicureanism as embracing speech act theory. Porter 1996: 622 discusses how "... in the search for a stable and naturally valid point of reference in language ..., Epicurus runs up against the empirical fact that *language is itself unstable, by nature and by convention*," thus making what is essentially a Derridean (if not exclusively Derridean) point about the search for any *archê* of language. Holmes 2005a: 571–72 makes explicit reference to Austin: "Words both are and are not things.

For generations, the fragmentary state of the evidence for ancient Epicureanism has rendered the task of understanding its thinking about language difficult, yet the research of the last fifty years has made that task much easier. The availability of many of Epicurus' papyrological fragments in the edition of Graziano Arrighetti as well as new editions of some of these fragments by David Sedley, Simon Laursen, and others provide the present-day interpreter of Epicurus with a wealth of information that was not available, or was only partially available, to an earlier generation of scholars.[57] Moreover, the large amount of material currently emerging from the Philodemus Project provides a steady flow of new results and readings for the eager student of Epicureanism. The present discussion considers a selection of this evidence insofar as it bears on the problem of speech acts in Epicureanism; however, even with all the new advances of the last forty years, any presentation can only hope to tell a partial, that is, a partially informed, story.

Perhaps the best way into the current account is through what has been perceived as a central conflict in the study of Epicurean semantics.[58] David Glidden, in a series of articles that remain controversial, attempted to revise what was then (late 1970s, early 1980s) the *communis opinio* on Epicurean semantics.[59] In particular, in his piece entitled "Epicurean Semantics," Glidden argued that "Epicurean linguistics shows little concern for how words acquire meanings and thereby describe what is happening in the world. Rather, Epicurus is really interested in how acts of speech can come to be

---

Lucretius goes to great lengths to prove that the voice is corporeal because it does things . . . but we are more in the world of J. L. Austin here than the world of body and void."

57. The up-to-date details of the various editions of the Herculaneum papyri (including Epicurus *On Nature*) may be found at http://www.herculaneum.ox.ac.uk/?q=books (Accessed 14 March 2013). As regards papyri, the present discussion, due to its focus on language, will draw most extensively upon David Sedley's edition of *On Nature* 28 (Sedley 1973), supplementing that text with other evidence as necessary.

58. The bibliography on Epicurean linguistic thinking is vast (and rapidly growing). Modern treatments include Shrijvers 1970, Long 1971, Manuwald 1972, Sedley 1973, Striker 1974, Goldschmidt 1978, Asmis 1984: 24–34, Glidden 1983 and 1985, Barnes 1996, Porter 1996, Holmes 2005a, Reinhardt 2008, and Atherton 2005 and 2009. Morel 2009: 140–51 is not focused purely on language, but he provides an insightful treatment of *prolepsis*, the Epicurean term for a concept or idea that figures importantly in our linguistic evidence. Beer 2009, although focused primarily on resonances between Philodemus and Lucretius, discusses much of the evidence for Epicurean epistemology and linguistic theory.

59. Of primary relevance are Glidden 1985 and 1983a, although Glidden 1983b, 1981, and 1979 all present material of some relation to the project under discussion. Atherton 2009: 198 groups Glidden with those who hold an "extentionalist interpretation" of language in Epicurean thought. She characterizes this position as one where "a direct signifying relation

used as labels for physical states.... Epicurean linguistics is less a theory of language than a theory of behavior."[60] A central point in Glidden's treatment is his revision of the understanding of the Epicurean technical term *prolepsis* (Gk. πρόληψις).[61] He insists that instead of functioning as linguistic meanings or concepts, *prolepseis*—while still allowing some generality—"designate some portion of reality," as is the case with Quinean natural kinds like man.[62] The position he was arguing against, one largely, if not exclusively, developed by A. A. Long and David Sedley in the early 1970s, holds that *prolepseis* are "*concepts* formed from repeated 'clear views' of the same thing or the same kind of thing,"[63] which combine a "simple semantic function" with "a central role in all acts of recognition performed by the human mind."[64]

To analyze the role of speech acts in Epicureanism, we must realize that both of these accounts get at part of the truth. They both, that is, capture aspects preserved in our evidence on Epicurean linguistic thinking. Moreover, they are not as opposed as it may seem.[65] Glidden, on the one hand, who denies Epicurus any conceptual abstraction whatsoever, correctly places a strong emphasis on the role that nomination and deixis play in Epicurean evidence, particularly accounts of the origins of language. By contrast, the traditional account better explains the way in which *prolepseis* function to allow abstraction, even if that abstraction seems, or in fact is, unjustified on a fully materialist worldview.[66] What I want to argue for is an Epicurus who

---

holds between language and things in the world." For similar views, see De Lacy 1939: 85, De Lacy 1978: 184, Asmis 1984: 26–7, and Annas 1992: 167.

60. Glidden 1983a: 187.

61. I use the transliterated form of this word unless it is contained in a fuller textual citation. Scholars generally assume (based upon Cicero *ND* 1.44) that Epicurus coined the term.

62. Glidden (1983a: 222n122) points out further correspondences between Epicurus and Quine: "Like Quine's classes and natural kinds, Epicurus allows himself both individuals (individual atoms, individual atomic conglomerates) and individual types of structures (round atoms, atomic clusters composing, say, water) in his ontology."

63. Long 1971: 119. Emphasis added. Morel 2009: 139 writes similarly: "il semble même que ce mot [*sc. prolepsis*] ... soit un terme générique qui envelope d'autres termes désignant des notions abstraites ou des opérations mentales ..."

64. Sedley 1973: 14.

65. In their stated forms, the two accounts are obviously opposed, but I want to suggest that it is possible to usefully retain aspects of both accounts.

66. It is also true that the vast majority of our source material uses (often Stoic) conceptual language. The definition of *prolepsis* at Diogenes Laertius 10.33, for example, reads more like a list of Stoic technical terms than a lucid definition. (Of course, definitions, and dialectic more generally, are not particularly Epicurean.)

looks more like J. L. Austin (and therefore like a less radical Derrida).[67] Just as Austin pits himself against an understanding of language as simply describing the world, as only making statements that are either true or false, so evidence within the remains of Epicurus' writing suggests that he was not simply interested in describing the world, although *prolepseis* (and the other criteria of truth) could ideally allow language that function as well. Indeed, we have already noted that a therapeutic philosophy would have great interest in acting upon the world; and there are a number of texts that support the idea that Epicurus was aware, at least in certain respects, of the implications of locating a theory of language within a theory of action.[68]

## (ii) Isolating the Performative

The case begins with two pieces of evidence, which Glidden has taken as giving central insight into the way in which all language functions, but which I would like to claim only describe the way in which the performative aspect of language functions. Both pieces of evidence, the first from Sextus, the second from Plutarch, suggest that Epicureans did not believe in a "signified" (τὸ σημαινόμενον), a term which—although more or less clearly marking Stoic influence in both sources—must designate some form of mental conception, an intermediate between a "signifier" (τὸ σημαῖνον) and a "referent" (τὸ τυγχάνον).[69] In the first, Sextus reports a difference of opinion between the Stoics and Epicureans on the question of whether there is something true. On the one hand, the Stoics posit that the true (or the false) lies in the

67. By "less radical," I mean only that Epicurus did not—although he rejected the methodologies of his philosophical contemporaries—engage in the kind of wordplay and formal experimentation characteristic of Derrida. (Epicurus was, however, more of a stylist than some imagine. What we know of his output, which even in its present fragmentary state ranges from aphorisms to lengthy treatises to letters, is quite varied and stylistically aware. Seneca, for one, seems to have taken him as something of a literary forebear. See Henderson 2004: 15 and *passim*. A traditional evaluation of Epicurus' style as artless is Norden 1898: 1.123.)

68. Speculation that an Epicurean theory of language belongs (at least in some sense) within a theory of action is supported by the remark of Proclus (*In Plat. Crat.* 17.4–7, p. 8 Pasquali): ὁ γὰρ Ἐπίκουρος ἔλεγεν ὅτι οὐχὶ ἐπιστημόνως οὗτοι ἔθεντο τὰ ὀνόματα, ἀλλὰ φυσικῶς κινούμενοι ὡς οἱ βήσσοντες καὶ πταίροντες καὶ μυκώμενοι καὶ ὑλακτοῦντες καὶ στενάζοντες. ("Epicurus used to say that names were not coined knowingly but by natural movements like coughing, sneezing, lowing, barking, and groaning.") Proclus' list is eclectic, including items we consider involuntary as well as ones we consider (at least somewhat) voluntary. Yet all these items are easily considered actions rather than (e.g.) simply descriptive speech.

69. This Stoic technical term is sometimes also translated as "name-bearer." (LS) Cf. the discussion at LS: 1.201 and Frede 1978: 32. These works derive the term from "that which obtains

signified (περὶ τῷ σημαινομένῳ); the Epicureans, by contrast, locate the true in the utterance (περὶ τῇ φωνῇ).[70] The Stoics maintain that there are three parts yoked together in an act of speech: the signifier (τὸ σημαῖνον), the signified (τὸ σημαινόμενον), and the referent (τὸ τυγχάνον);

> οἱ δὲ περὶ τὸν Ἐπίκουρον καὶ Στράτωνα τὸν φυσικὸν δύο μόνον ἀπολείποντες, σημαῖνόν τε καὶ τυγχανόν, φαίνονται τῆς δευτέρας ἔχεσθαι στάσεως καὶ περὶ τῇ φωνῇ τὸ ἀληθὲς καὶ ψεῦδος ἀπολείπειν. (Sextus Empiricus *Adv. Math.* 8.13)
>
> but the followers of Epicurus and Strato, the student of nature (τὸν φυσικὸν), leave only two parts, the signifier and the referent, and they seem to belong to the second faction, who leave the true and false in the utterance.

Plutarch, in the second passage, sounds a similar note:

> τίνες μᾶλλον ὑμῶν (*sc.* Ἐπικούρειοι) πλημμελοῦσι περὶ τὴν διάλεκτον, οἳ τὸ τῶν λεκτῶν γένος οὐσίαν τῷ λόγῳ παρέχον ἄρδην ἀναιρεῖτε, τὰς φωνὰς καὶ τὰ τυγχάνοντα μόνον ἀπολιπόντες, τὰ δὲ μεταξὺ σημαινόμενα πράγματα, δι᾽ ὧν γίνονται μαθήσεις διδασκαλίαι προλήψεις νοήσεις ὁρμαὶ συγκαταθέσεις, τὸ παράπαν οὐδὲ εἶναι λέγοντες; (Plutarch *Adv. Col.* 22, p. 1119f = Usener 259, part (Usener 1887: 189))
>
> Who strike the wrong note more than you (*sc.* Epicureans) with regard to dialectic, you who entirely destroy the class of *lekta* ("sayables"), which provide substance to speech, and leave only sounds and their referents, instead asserting that the intermediate things—the signifieds through which learning, teaching, preconceptions, thoughts, impulses, assent come to be—do not exist at all?

As Plutarch makes clear in his Stoicizing terminology, getting rid of signifieds (τὰ σημαινόμενα) seems to destroy the possibility of *prolepseis*, at least on

---

a case" (τὸ τυγχάνον πτώσεως *vel sim.*), and it is generally agreed that it refers to "the external objects that have the qualities signified." (Frede) Asmis 1984: 26, by contrast, while maintaining that the term designates the "underlying external thing," nonetheless suggests a different derivation when she remarks that it is literally "what happens." "Referent" is perhaps the best modern equivalent of the term; its connotations are, however, clearly different, and Saussurean parallels should only be carried so far.

70. On the notion of truth employed here ("truth in the utterance"), see Cicero *De Fato* 37.

their traditional interpretation. For this reason, some have seen fit to ignore these two reports. The primary exception to this generalization, Glidden, was compelled to construct an alternative notion of *prolepsis*, as noted earlier.[71] Between ignoring this evidence and using it as the primary basis for constructing Epicurean linguistic thought, however, there is a middle course—reading these reports as a description of performative language.

Such an interpretation requires the presumption that Sextus and Plutarch were mildly confused, but it has the advantage of using this evidence in a manner that coincides strikingly with modern descriptions of a performative utterance.[72] Consider the words of linguist Émile Benveniste:

> The performative utterance, as it is an act, has this property: it is *unique*. It cannot occur except in particular circumstances, once and once alone, at a specific time and place. It has neither the value of a description nor of a prescription but ... of an accomplished act. This is why it is often accompanied by indications of date, place, personal names, witnesses, etc.; in short, it is an event because it creates the event. As an individual act, as a fact of history, a performative utterance cannot be repeated. Each reproduction is a new act, which the one with the appropriate power accomplishes ...
>
> This analysis brings us to the realization that the performative has a singular property: it is *self-referential*; it makes reference to a reality that it establishes itself; it is an effective utterance under those conditions which make it an act. Whence it arises that it is at once a linguistic occurrence—since it must be pronounced—and a fact of reality, in so far as it is an accomplished act. The act, therefore, is identified with

71. Barnes 1996: 209–10 (following Striker 1974: 71–72) also offers a slightly different notion of *prolepseis*, at least in so far as he feels "concepts or preconceptions must possess or admit a propositional structure" (209). He bases his assertion on the claim that "the sorts of items which can be proved ... are propositional items" but provides no further ancient evidence to buttress this claim. What we know of thinking in Epicureanism suggests that it is not propositional but imagistic. See *DRN* 4.722–817, Annas 1992: 165, Atherton 2009: 204, and Shearin 2014: 192.

72. Or rather: these two reports are "confused" in the sense that they seem to take a linguistic description of a performative utterance as a paradigm for all language; however, if we hold with Derrida (and some writings of Austin), that any separation of performative from constative is only provisional (and doomed to failure), then this error may not be so grave, consisting only in giving a partial, idealized description of the powers of language rather than in extending the description of one type of language to all language. It also seems possible to claim that while Plutarch is undoubtedly somewhat confused, Sextus is perfectly correct, although he is comparing two totally different notions of truth, one that understands a statement as a true description (the Stoic version), a second that understands a statement as a true act (the Epicurean).

> the statement of the act. The signified is identical to the referent. This is that to which the phrase "With these words ..." bears witness. The utterance that is itself taken for reference is indeed self-referential.[73]

It does not take a great effort of imagination to understand how this description fits with those provided by Sextus and Plutarch. There is no "signified" in the Epicurean analysis reported by both authors because the "signified" is, on Benveniste's view, "identical to the referent." τὸ σημαινόμενον and τὸ τυγχάνον, in other words, are one and the same. The uniqueness of the performative utterance guarantees that there can be no *prolepsis*, no empirical preconception, of any performative as an individual act.[74] (For the standard example of the marriage act, this analysis means that the marriage rite, with its performative "I do" or "I will," creates the marriage to which it refers. Further, the marriage is necessarily unique. Although the words of the rite are formulaic, the individual actors and the individual event, which make the rite unique, are the only reason the rite is carried out in the first place.) Reading these two wayward pieces of evidence in the above fashion provides an impetus for investigating whether there are further traces of speech act theory within surviving Epicurea. What we find if we undertake such an investigation, particularly in Epicurus' thinking about logic and dialectic, is quite striking.

## (iii) Dialectic and Authority

Most important are the various remains to be found in Cicero, chiefly but not exclusively in his *De Fato*. Here we read (in more than one place) that Epicurus, just like Austin, "denies that every assertion is either true or false" (*De Fato* 21: ... *negare omnem enuntiationem aut veram esse aut falsam* ...). Cicero brings out this point in mockery of Epicurus' logical prowess, but perhaps, as with Austin, such a claim is merely part and parcel of an Epicurean discovery of the performative utterance:

> necesse est enim in rebus contrariis duabus (contraria autem hoc loco ea dico, quorum alterum ait quid, alterum negat)—ex iis igitur necesse

73. Benveniste 1966: 273–74 (emphasis in French original). Cf. Agamben 2011: 55, who comments that the self-referential character of the performative expression "is not exhausted in the fact that the performative ... takes itself as referent. Rather it is necessary to specify that the self-referentiality of the performative is constituted always by means of a suspension of the normal denotative character of language."

74. To be clear: one may have a *prolepsis* of marriage generally but not of *this* marriage.

est invito Epicuro alterum verum esse, alterum falsum, ut "sauciabitur Philocteta" omnibus ante saeculis verum fuit, "non sauciabitur" falsum; nisi volumus Epicureorum opinionem sequi, qui tales enuntiationes nec veras nec falsas esse dicunt, aut cum id pudet illud tamen dicunt, quod est inpudentius, veras esse ex contrariis diiunctiones, sed quae in his enuntiata essent eorum neutrum esse verum. o admirabilem licentiam et miserabilem inscientiam disserendi! si enim aliud in eloquendo nec verum nec falsum est, certe id verum non est; quod autem verum non est qui potest non falsum esse? aut quod falsum non est qui potest non verum esse? (Cicero *De Fato* 37–38)

It is necessary in the case of two contrary matters (here I call "contrary" these things for which the one affirms something, the other denies it)—of contraries, then, it is necessary, though Epicurus is unwilling to admit it, that one is true, the other false, as (e.g.) "sauciabitur Philocteta" ("Philoctetes will be wounded") was true in all previous ages, "non sauciabitur" ("Philoctetes will not be wounded") was false; unless we want to follow the opinion of the Epicureans, who assert that such assertions are neither true nor false, or when this causes shame, they nevertheless state the following (which is more shameful): that disjunctions from contraries are true, but neither of the things which have been asserted in these contraries is true. O wondrous license and pitious ignorance of discourse! If, in speaking, the former statement is neither true nor false, surely the latter is not true; but how is it possible that what is not true is not false? Or how can what is not false be untrue?

In ridiculing Epicurus, Cicero preserves important clues as to why Epicurus (and Epicureans) may have resisted the (rather Stoic) insistence that all "disjunctive contraries" are either true or false. If, as we have been suggesting, Epicurus resists on the grounds that he has uncovered the Austinian performative, then he denies that certain assertions are true or false on the grounds that they are not, or cannot be, purely constative. This surmise is perhaps borne out by the fact that Epicurus was notoriously reluctant to agree that statements in the future tense were true or false: future statements cannot be purely descriptive or "constative," at least not in the sense that they describe some state of affairs which has already been observed (or is currently being observed) by the speaker.[75]

75. For the clear emphasis on future "disjunctive contraries," see Cicero *ND* 1.70 and *Lucullus* 97–98, both discussed in this chapter. See, too, Warren 2006. From the *Lucullus* passage, it

Moreover, the fact that Epicurus' followers assert, laughably in Cicero's eyes, a double explanation of their position, at first denying, like Epicurus, that such utterances are true or false, but then also offering a further, alternative explanation—that "disjunctive contraries" are true, but "neither of the things which have been asserted in these contraries is true" (*quae in his enuntiata essent eorum neutrum esse verum*)—finds parallel within Austin's extant corpus. The transcript of an exchange at a 1958 colloquium between J. L. Austin and a French philosopher identified as M(ons). Devaux offers the following:

AUSTIN: A performative utterance cannot be true or false. When I say to you, "I bet you five francs that it will rain tomorrow," a sentence perfectly natural and ordinary, the bet which I make cannot be qualified by true or false. What may be true or false is the fact that I bet. I could have bet you the opposite, or I could have not bet at all.

DEVAUX: It can be true or false in two senses: true or false with regards to the speaker; true or false in the sense of the bet....

AUSTIN: But let's begin again: "Close the door!"—can this be true or false?

DEVAUX: Yes, as an act stated in an utterance.[76]

Devaux and Austin never resolve their dispute. In the transcript, Jean Wahl intervenes, and this branch of discussion is abandoned. Nevertheless, even in their undeveloped state, these remarks may permit us to glimpse how the followers of Epicurus allowed in their second explanation not only that "disjunctive contraries" (*diiunctiones ex contrariis*) were true, but also that neither half of the disjunction was true: the Epicureans may have allowed that the disjunctive utterance was true qua act. This explanation fits well with Epicurean epistemology more generally, which affirmed "the truth of all impressions," presumably including the truth of the perception of an utterance.[77] Yet such an

is clear that Epicurus' denial applied to all "disjunctive contraries." In this context, it is worth recalling Epicurus' words on the future in the *Letter to Menoeceus* (127): μνημονευτέον δὲ ὡς τὸ μέλλον οὔτε ἡμέτερον οὔτε πάντως οὐχ ἡμέτερον, ἵνα μήτε πάντως προσμένωμεν ὡς ἐσόμενον μήτε ἀπελπίζωμεν ὡς πάντως οὐκ ἐσόμενον. ("We must remember that the future is neither ours nor wholly not ours so that we neither fully anticipate that it will occur nor utterly despair that it will not occur.") Such statements make clear the point that Epicurus felt the future to be something about which we did not have secure empirical knowledge.

76. Austin 1962b: 289–91.

77. Basic texts and discussion of this central point in Epicurean epistemology may be found at LS: 1.78–86 and in the corresponding section of the second volume, 2.83–91. Particularly

allowance—much like M. Devaux's challenge to Austin—in no way changes the fact (Austin's main point) that the content of a performative utterance (at least in its paradigmatic form) cannot, despite Cicero's frustration, be qualified as true or false, if by "true" or "false" we mean, respectively, a correct or incorrect description of some state of affairs. In other words, allowing two different senses of "true" may enable the Epicureans both to deny and to affirm the truth of "disjunctive contraries."

The *De Fato* passage finds multiple parallels within the Ciceronian corpus, notably in the somewhat more charitable version of the *Lucullus*:[78]

> etenim cum ab Epicuro, qui totam dialecticam et contemnit et inridet, non inpetrent ut verum esse concedat quod ita effabimur "aut vivet cras Hermarchus aut non vivet," cum dialectici *sic* statuant, omne quod ita disiunctum sit quasi "aut etiam aut non" <non> modo verum esse sed etiam necessarium (vide quam sit catus[79] is quem isti tardum putant: "si enim" inquit "alterutrum concessero necessarium esse, necesse erit cras Hermarchum aut vivere aut non vivere; nulla autem est in natura rerum talis necessitas")—cum hoc igitur dialectici pugnent, id est Antiochus et Stoici; totam enim evertit dialecticam: nam si e contrariis disiunctio (contraria autem ea dico, cum alterum aiat alterum neget)—si talis disiunctio falsa potest esse, nulla vera est. (Cicero *Lucullus* 97–8)

---

pithy and relevant is the remark of Diogenes Laertius 10.32: καὶ τὸ τὰ ἐπαισθήματα δ' ὑφεστάναι πιστοῦται τὴν τῶν αἰσθήσεων ἀλήθειαν. ("And the establishment of perceptions lends belief to the truth of [all] sensations.")

78. Cf. Plutarch *De Pyth. orac.* 398F, where the Epicurean-leaning spokesman Boethus comments: τί δ' ἔστι τῶν ἀτόπων καὶ ἀπροσδοκήτων περὶ γῆν ἢ θάλατταν ἢ πόλεις ἢ ἄνδρας, ὃ τις ἂν προειπὼν οὐ τύχοι γενομένου; καίτοι τοῦτο γε σχεδὸν οὐδὲ προειπεῖν ἔστιν ἀλλ' εἰπεῖν, μᾶλλον δὲ ῥῖψαι καὶ διασπεῖραι λόγους οὐκ ἔχοντας ἀρχὴν εἰς τὸ ἄπειρον· οἷς πλανωμένοις ἀπήντησε πολλάκις ἡ τύχη καὶ συνέπεσεν αὐτομάτως. διαφέρει γὰρ οἶμαι γενέσθαι τὸ ῥηθὲν ἢ ῥηθῆναι τὸ γενησόμενον. ("What strange, unexpected thing is there concerning earth or sea or cities or men, which anyone, after forecasting it, could not happen to have come to pass? And yet *this* is not, I dare say, prophecy [προειπεῖν] but mere speech [εἰπεῖν]—or rather casting about and sowing words lacking authority [ἀρχὴν] *ad infinitum*. Many times has chance met with wandering words and brought them to pass by accident. What was said coming to pass differs, I believe, from what will come to pass being said.") Warren 2006: 381 reads the continuation of this passage (399A) as "at odds" with *De Fato* 37–38, but the chief point (which is clear in the extract I have cited) is that προειπεῖν and εἰπεῖν are not the same, which need not be at odds with the refusal to accept that all bivalences are true or false. Indeed, it seems likely that the difference between the two forms of speech is one of authority (and knowledge): only the speaker of προειπεῖν has the authority (in Plutarch, ἀρχή) to make prophecy felicitous.

79. catus Lambinus| cautus Mss. Apart from this emendation (accepted in Usener 1887), the text here follows Plasberg (ed.) 1922. Denys Lambin's *catus* gives the statement greater point. The transmitted *cautus* provides a much weaker contrast with *tardus*.

> And I say this since they (*sc.* the dialecticians) were not able to get Epicurus, who scorns and mocks all dialectic, to admit the truth of what we shall proclaim (*effabimur*) thus: "aut vivet cras Hermarchus aut non vivet" ("either Hermarchus will live tomorrow or he won't"), since the dialecticians put it thus: every disjunctive statement of this sort "aut etiam aut non" ("either A or not-A") is not only true but also necessary. (See how clever Epicurus is, whom *those men* think slow: "If I grant," he says, "that one of the two possibilities is necessary, it will be necessary tomorrow that Hermachus either be alive or not be alive; but there is no such necessity in the nature of things.") With this man, then, they quarrel, the dialecticians, that is, Antiochus and the Stoics; for he upsets all dialectic: for if a disjunction of contraries (and by "contraries" I mean, when the disjunction affirms one possibility, denies the other)—if such a disjunction can be false, no disjunction is true.

This passage, while re-visiting the same topic as the earlier one, introduces a number of subtle yet important variations: Epicurus is no longer a foolish failure at logic but a clever student of language, one who subverts all dialectic. More important for the study of the Epicurean performative, we read here an actual (if doubtless fabricated) defense of Epicurus' position on disjunctive, "contrary" statements (again, paradigmatically, if not exclusively, future statements): Epicurus says, "If I grant (*concessero*) that one of the two possibilities is necessary, it will be necessary tomorrow that Hermachus either be alive or not be alive; but there is no such necessity in the nature of things." What is so striking is that Epicurus seems to be speaking not simply about Hermarchus and his possibilities as a physical agent in the world (with the disjunctive statement as a pure description of those possibilities) but about the impact of his own speech upon the world. If we attend closely to the words placed into Epicurus' mouth by Cicero, he proclaims that his own verbal concession could constitute necessity in nature, a necessity that is not otherwise present.[80] Such an interpretation certainly reads Epicurus' speech within a theory of action, allowing Epicurus to

80. Note the close verbal parallels at Cicero *ND* 1.70: *idem facit contra dialecticos; a quibus cum traditum sit in omnibus diiunctionibus, in quibus 'aut etiam aut non' poneretur, alterum utrum esse verum, pertimuit ne, si concessum esset huius modi aliquid "aut vivet cras aut non vivet Epicurus," alterutrum fieret necessarium: totum hoc "aut etiam aut non" negavit esse necessarium; quo quid dici potuit obtusius?* ["He does the same thing (*sc.* says something more disgraceful than the admission that he cannot argumentatively defend his position) in response to the dialecticians: when it was stated by them that in all disjunctive statements, in which "aut etiam aut non" ("either A or not-A") appears, one of the two possibilities is true, he feared that, if something of this sort were granted, e.g, "aut vivet cras aut non vivet Epicurus" ("either Epicurus will live

utter a "magic" performative, but it also calls attention to an issue often raised in discussions of speech acts, namely, the authority of the speaker.[81]

Concerning the authority of speakers of performative utterances, Émile Benveniste has remarked:

> ... a performative utterance only possesses reality if it is authenticated as an *act*. Beyond the circumstances which render it performative, such an utterance is nothing. Anyone can cry out in a public space: "I decree a movement of the army." As it cannot be an *act* without the requisite authority, such a pronouncement is nothing more that words....A performative utterance which is not an act does not exist. It exists only as an act of authority.[82]

The consideration of authority thus returns us to a problematic upon which we lingered for some time in a more abstract mode—that of context. The speaker's authority is one of the circumstances that render the performative performative. (We may recall Austin's example of appointing when not entitled to appoint.) The very diction of the passage in fact suggests that such consideration, both of the performative and of the authority that renders language performative, is not out of place: *effabimur* ("we shall proclaim"), the verb employed to introduce the disjunctive statement (whose necessity Epicurus must decide), is, as Maurizio Bettini has noted, a rather special verb:

> ... the word [sc. *effari*] held a place of absolute importance in religious language. (*de Oratore* 3.153). In fact, the expression *effatus* formed part of a technical religious vocabulary indicating a place that had been

---

tomorrow or he will not"), one of the two possibilities would become necessary: he entirely denied that this "aut etiam aut non" ("either A or not-A") was necessary; what could be uttered more thick-headed than that?"] Compare, too, the language at Cicero *Fin.* 1.25.

81. It would be remiss not to mention a fragment of Epicurus *On Nature*, Book 25 (Arr.[2] 34.26–27) that seems much in line with the points made in Cicero's *De Fato*. The relevant lines (24–30) run something like the following: ... νουθε[τ]εῖν τε ἀλλήλους καὶ μάχε[σ]θαι καὶ μεταρυθμίζειν ὡς ἔχοντας καὶ ἐν ἑαυτοῖς τὴν αἰτίαν καὶ οὐχὶ ἐν τῆι ἐξ ἀρχῆς μόνον συστάσει καὶ ἐν τῆι τοῦ περιέχοντες καὶ ἐπεισιόντος κατὰ τὸ αὐτόματον ἀνάγκηι. ("to rebuke, oppose and reform each other as if the responsibility lay also in themselves, and not just in their congenital make-up and in the accidental necessity of that which surrounds and penetrates them.") (Trans. Sedley) What is so striking here is that, as with Epicurus' apparent belief that his concession may create (physical) necessity, speech in the form of admonishment seems able to *do* things. For the latest edition of this papyrus, see Laursen 1995 and 1997, although new readings and emendations have been proposed since that edition.

82. Benveniste 1966: 273. Emphasis in French original.

> "defined" with the "sure words" (*certa verba*) of ritual: a field in which *auguria* were going to be taken, a temple, the *pomerium* itself.[83] Varro, describing the way in which the augurs marked out spaces for observing celestial signs, says: *hinc effata dicuntur, qui augures finem auspiciorum caelestum extra urbem agri<s> sunt effati ut esset; hinc effari templa dicuntur: ab auguribus effantur qui in his fines sunt,* "From this (i.e., from *fari*) comes the expression 'spoken things' (*effata*), referring to the formulas[84] with which[85] the augurs have marked out the boundaries of the fields, outside the city, for observing celestial omens. From this, areas marked out for augury are also said to be 'defined with a word' (*effari*), and the augurs declare (*effantur*) their boundaries" (*de Ling. Lat.* 6.53).... *Effari* is a mode of speaking that not only declares the sacred nature of a space, but it directly *realizes* that nature ...[86]

While it would be difficult to presume that *effabimur* in *Lucullus* is being used in the special, technical sense that Bettini and Varro describe, such evidence nonetheless indicates that the word was standardly used performatively. Cicero himself notes elsewhere (*De oratore* 3.153–54) that the use of *effari* renders a speech "loftier and more archaic" (*grandior atque antiquior*), so any thorough understanding of the *Lucullus* passage must strive to explain its presence. Whatever else it may do, *effabimur* once again raises the question of the authority of the speakers in the passage: as Epicurus' words seem to indicate a belief that his speech may act, creating necessity, so *effabimur* raises the possibility that the dialectic presented in *Lucullus* is not a contest of logic

83. Bettini's note: "Varro *de Ling. Lat.* 6.53; Cicero *ad Att.* 13.42, *de Leg.* 2.8.20; Gellius *NA* 13.14; Servius in *Aen.* 1.446, 2.692, 6.197, etc. On the *certa verba* necessary for an act of *effari,* Cicero also gives us an interesting bit of evidence. As is well known, during the exile to which he was condemned by Clodius's machinations, his house had been razed and the ground on which it stood had been declared consecrated. But Cicero considered that this had been done not only unfairly, but also illegally: *quid mirum si iste [metus] furore instinctus, scelere praeceps, neque institutas caerimonias persequi neque verbum ullum sollemne potuit effari?* ('What wonder is there if he, inspired by anger and driven by crime, was unable to complete the established rituals or utter [*effari*] even a single solemn word?' [*de Domo* 141]). In this case, too, *effari* presupposes an utterance of solemn, ritual words, sanctioned by tradition and law."

84. Bettini's note: "Servius in *Aen.* 6.197 explains *effata* with *preces: proprie effata sunt augurum preces: unde ager post pomoeria, ubi captabantur auguria, dicebatur effatus* ('Properly, *effata* are the prayers of the augurs. For this reason, a field outside the *pomerium* where *auguria* were taken was said to be *effatus* [spoken].')"

85. Bettini's note: "The *qui* of Varro's text is instrumental: cf. Norden 1995: 32n1."

86. Bettini 2008: 18–19. Emphasis original.

but a contest of authority, a contest not of who can better describe the truth but of who is authorized to create it.[87]

To make clearer what is at stake, it will be useful to describe the semantics (and pragmatics) of *fari* in somewhat greater detail: *fari*, it should be noted, is not just the root of *effari* but also at the root of *fatum*, the topic of discussion in Cicero's *De Fato*, where (as we have already seen) so many of the references to Epicurus' logic are found.[88] One example will indicate some of the semantic (and pragmatic) sediment lodged in the term: Statius, in describing the words of Jupiter (*Thebaid* 1.212–13), remarks *grave et immutabile sanctis / pondus adest verbis, et voces fata sequuntur* ("A heavy, immutable weight attends his holy words, and fate (*fata*) follows his utterances"). In discussing this passage, Bettini remarks, "*Fari* is . . . an absolute assertion that cannot be contradicted because it is not limited to the act of speaking: it simultaneously actualizes the utterance."[89] In other words, at least in this example, *fari* names the ideal (but hardly any actual) performative utterance, speech that creates in being spoken.[90] That Jupiter is the speaker here is, of course, hardly irrelevant to the present question of authority. Indeed, if Bettini's study of the term is correct, divine authority is always at play in any act of *fari*:

> The notion of "agency" in linguistic anthropology, meaning the capacity to perform social actions or to control how they are conducted, may help clarify the situation in which a speaker of *fari* finds himself.[91] In the sphere of *fari*—i.e., the mode of speech in which a prophet reveals hidden truths; a praetor makes legally-binding utterances; a priest recites effective ritual formulas; or an augur creates, with his words, a sacred

---

87. The authority of Epicurus' speech seems to extend also to the authoritative role his texts played within his school. See Fletcher 2012 for a discussion of the performative authority of the collection of Epicurean maxims, the *Kuriai Doxai*, within the Second Sophistic.

88. There is a growing amount of modern philosphical writing that puzzles over why Epicurus (and others in antiquity) would subscribe to what is called "the interentailment thesis," namely, that "logical determinism" (the claim that all statements must be true or false) implies "causal determinism" (one definition of "fate"). This literature, which I will not engage here, seems to me plagued by a failure to consider the semantics (and pragmatics) of the word *fari*. (Or, in the case of Greek terms for "fate" [chiefly εἱμαρμένη], such discussions fail to consider language as a part of material reality subject to "causes.") For one version of such a discussion, see O'Keefe 2005: 138–44.

89. Bettini 2008: 315.

90. It may even be prudent not to call such an utterance a "performative," as it is not subject to infelicity and other conditions of the performative. See Miller 1995: 156–60.

91. On agency in linguistic anthropology, see Ahearn 2001.

> space—it is clear that there is not one agency alone at work (as normally occurs in utterances) but two: that of the "animator" (the immediate speaker) who actually utters the message and that of the "principal," who simultaneously guarantees its authenticity and its efficacy.[92]

Such an understanding of *fari* casts an interesting light on the discussions of both *De Fato* and *Lucullus*, where the standoff may come down to which of the two sets of "animators" (to adopt Bettini's terminology), the Epicureans or their (seemingly) Stoic opponents, has the "principal" to render its speech efficacious. On this understanding, both Epicurus and his opponents are trying to appropriate traditionally religious language (as in Austin's studies the language of marriages, oaths, etc., is traditionally religious), specifically the (perceived) efficacy and authority of traditional, religious language.

Cleanthes' *Hymn to Zeus*, to choose but one well-known example, proves that the Stoics were adept at deploying traditional religious imagery and language to their own ends.[93] The Epicureans, by contrast, often have been portrayed as closet-atheists, who were unwilling to admit their atheism only out of fear of popular backlash. Yet the importance of religious language (and authority) for speech acts, for any notion of performative language, may help to explain the presence of comments like the following within Epicurus' corpus:

> Παρρησίᾳ γὰρ ἔγωγε χρώμενος φυσιολογῶ<ν> χρησμῳδεῖ<ν> τὰ συμφέροντα πᾶσιν ἀνθρώποις μᾶλλον ἂν βουλοίμην, κἂν μηδεὶς μέλλῃ συνήσειν, ἢ συγκατατιθέμενος ταῖς δόξαις καρποῦσθαι τὸν πυκνὸν παραπίπτοντα παρὰ τῶν πολλῶν ἔπαινον. (Epicurus *VS* 29)

> Frankly speaking, I would rather choose, in investigating natural phenomena, to deliver benefit to all men prophetically—even if no one is going to understand—than reap the vast praise falling from the mouths of the masses by agreeing with common opinion.[94]

---

92. Bettini 2008: 336.

93. See Asmis 1982, which argues for the influence of Stoic views of Zeus on Lucretius' opening hymn to Venus. This essay discusses Cleanthes' hymn extensively.

94. See Diskin Clay (1983a: 49), who remarks that Epicurus "called his philosophy prophecy." More recently, Michael Erler asserts (2009: 62) "epikureischer Philosophie geht es darum, Mantik als Element traditioneller Religiosität zu desavouieren, weil sie den erhofften therapeutischen Effekt für richtige Lebensfüherung nicht einlösen kann, ... Mantik wird ... durch

Here, although one function of the mention of prophetic speech is undoubtedly to conjure up notions of obscurity ("no one is going to understand"), the invocation of prophecy may also refer to the authority of prophetic speech, which is at once suspect and divine. Indeed, prophetic speech is simply one type of performative utterance, structurally a type of promise, whose function is historically (if not in modern times) guaranteed by an absent divinity. Epicureanism, with a conception of divinity as not only absent but also as failing to intervene in the world, seemingly has a weak claim on any prophetic authority, on any "principal"; but it may be precisely with an interest in revising such assessments that it persists in its engagement with religious ideas and language.[95]

It is at this point that we need to invoke the Derridean notion of context, which an earlier section of this chapter described. Once again in the formulation of J. Hillis Miller: "[t]he context is there already, but it becomes a context only when the speech act intervenes within it, however weakly and without power to saturate it. The speech act nevertheless transforms the context it enters, even though in retrospect that context seems to have been there already as the ground of the speech act's efficacy." What is so important about this definition is that it points to the fact that neither speech acts nor their contexts are fixed; and yet they seem to be so. That it is so standard and formulaic (both linguistically and otherwise) is what makes a marriage rite a recognized and recognizable ritual; and yet each performance of the rite is necessarily unique, and it may in fact augment and modify the standard ritual, although it is arguably impossible to know when or whether one has substantively modified it.[96] Epicurus, if only implicitly, recognized not only the power of the performative but also this more difficult point about the role of the context in performative utterances.

This line of explanation would render comprehensible charges against Epicurus like the following:

> ὑποκρίνεται γὰρ εὐχὰς καὶ προσκυνήσεις οὐθὲν δεόμενος διὰ φόβον τῶν πολλῶν καὶ φθέγγεται φωνὰς ἐναντίας οἷς φιλοσοφεῖ· καὶ θύων μὲν ὡς μαγείρῳ παρέστηκε τῷ ἱερεῖ σφάττοντι, θύσας δὲ ἄπεισι λέγων τὸ

---

Epikurs Philosophie und epikureische Texte ersetzt." Erler effectively contrasts Platonism and Stoicism (which circumscribe the role of prophecy) with Epicureanism (which presents itself as a new form of, or replacement for, prophecy).

95. The fact that Lucretius referred to Epicurus as a god (*DRN* 5.8: ... *deus ille fuit, deus* ...) is hardly irrelevant in this context.

96. It is of course possible to recognize that one has deviated from a given ritual; but it is difficult, perhaps impossible, to know when a ritual has changed qua ritual, even if that change

> Μενάνδρειον "ἔθυον οὐ προσέχουσιν οὐδέν μοι θεοῖς." (Plutarch *Contra Beat.* 1102b)[97]
>
> He plays along, performing prayers and worshipping, omitting nothing out of fear of the masses; and he pronounces utterances (φωνάς) contrary to his philosophical teachings. Moreover, at sacrifices he stands next to the slaughtering priest as if he were a cook; and after the sacrifice he goes away, uttering the saying of Menander: "I was sacrificing to the gods who fail to pay attention to me."

What (aside from the fear of popular opinion suggested by Plutarch) would inspire Epicurus to attend religious rituals and recite utterances contrary to his own teachings? A possible answer is that doing so provided him access to a storehouse of well-established performative utterances (and related ritual behaviors) and an ability, if one not easily controlled, to co-opt that performative power and the authority associated with it.[98] The line of Menander, at least on one reading, could suggest such a strategy: Epicurus sacrifices to the gods for backing up his authority; the gods do not intervene, just as he says they should not.[99] Epicurus did after all proclaim that, as the ultimate result of his philosophy, "you shall live as a god among men" (*Letter to Menoeceus* 135): such a pronouncement is rightly taken to refer to the ethical goal of Epicureanism, ἀταραξία ("freedom from disturbance"), but it may also designate the authority of the speech of someone who has attained such a goal.

## (iv) Performative-Constative

Concerns about religious language and authority and their role in performative language more generally will detain us at greater length in *De rerum*

---

may be perceptible after the fact. This point is essentially the same as de Man's point that "[p]erformative rhetoric and cognitive rhetoric, the rhetoric of tropes, fail to converge" (de Man 1979: 300).

97. Usener 31, part (Usener 1887: 103).

98. By "access," I mean here not merely that attending the ritual would provide Epicurus with knowledge of a given rite or sacrifice, but also that his attendence would be a necessary prerequisite for engaging in actual performative acts, perhaps with an eye to changing the operation of those acts.

99. "Not paying attention" is what Epicurean gods are supposed to do: see, e.g., Lucretius *DRN* 6.68–79, Cicero *ND* 1.45a.

*natura*.[100] At present, however, I want to move this section nearer to conclusion by suggesting ways in which, on the Epicurean account, the performative aspect of language fits in with its constative aspect. Epicurus surely also recognized the constative function of language, even if he believed that that function may never be fully separated from the performative one (and therefore dismissed dialectic). Indeed, much of the previous scholarly literature on Epicurean semantics, especially that which stakes a claim in the debate over *prolepseis*,[101] has been devoted to identifying and treating language as constative, so the aim of the present discussion is not to prove that language, on an Epicurean account, may be in some sense constative—that has already been shown amply—but rather to demonstrate that numerous pieces of traditional evidence may be treated as allowing space for both the performative and the constative.

Let us begin with a few smaller pieces of evidence:

> τὴν διαλεκτικὴν ὡς παρέλκουσαν ἀποδοκιμάζουσιν· ἀρκεῖν γὰρ τοὺς φυσικοὺς χωρεῖν κατὰ τοὺς τῶν πραγμάτων φθόγγους. (Diogenes Laertius 10.31)
>
> They dismiss dialectic as a diversion, for it is sufficient for students of nature to proceed in accordance with the utterances of things.

> τῶν τε ζητήσεων εἶναι τὰς μὲν περὶ τῶν πραγμάτων, τὰς δὲ περὶ ψιλὴν τὴν φωνήν. (Diogenes Laertius 10.34)
>
> Of inquiries, generally speaking, some are about things (περὶ τῶν πραγμάτων), others concern mere sound.

These two passages re-iterate Epicurus' disdain for dialectic, but they also introduce a vocabulary which has been less prominent in our discussions to

100. A fuller account may consider the close connection between the origins of language and the origins of religion in extant Epicurean accounts, a fact that undoubtedly has significance for performative language.

101. I offer some remarks on *prolepseis* below, but it is not my intention to have the final word in that debate. As to whether *prolepseis* are (more) predominantly features of the world or concepts in our head, it seems to me that *prolepseis* must, as they are empirically formed, be both part of the world *and* in our heads. Whether they are more concept-like or more (as Glidden [1985: 196] seems to think) like automatic behavioral responses does not seem to be a question contemplated by our sources; our evidence seems to prefer the language of concepts, whether that is the fault of Stoic influence or not. What interests me about *prolepseis*, though, is how they relate to a changing "context."

this point: both passages point to the fact that Epicurean analysis is interested in language (φϑόγγοι) concerning things (τὰ πϱάγματα). "Things" seems obviously to refer to the physicist's concern with physical reality, with atoms and void; but our consideration of Epicurus' dismissal of dialectic in Cicero provides another line of attack. πϱᾶγμα in Greek may also be an "act" (cf. LSJ[9] s.v. I): in fact, etymologically, it is the achieved result of an act of doing (πϱάττειν), which can quite easily be an act in the sense of a performative utterance. In making such a point, one would hardly wish to deny that Epicurus was interested in "things" in the traditional sense of physical reality; but his stated interests may already make room for the performative alongside the constative.[102] If such a conclusion is possible, it is also worth paying attention to the negative foils in the cited oppositions: in the first, dialectic is dismissed; in the second, the object of attack is "mere sound" (ψιλὴ φωνή). What do dialectic and "mere sound" share? *Dicta* such as the following may provide a clue: κενὸς ἐκείνου φιλοσόφου λόγος, ὑφ᾽ οὗ μηδὲν πάϑος ἀνϑϱώπου ϑεϱαπεύεται ("Empty is the word of that philospher by whom no human suffering is alleviated ...").[103] If κενός ("empty") here means something like ψιλή ("bald, unadorned"), then Epicurus' attack may be precisely on discourse that aims at being purely constative, or at least at discourse that fails to create therapeutic action, whether that be accomplished directly in speaking or indirectly through speaking. Whatever the case, bearing these smaller pieces of evidence—and the space they create for the performative alongside the constative—in mind, let us move on to lengthier passages.

"Empty utterances" (κενοὶ φϑόγγοι), which we just saw ridiculed, occur elsewhere in Epicurus' writings, notably in one of the most cited passages of the *Letter to Herodotus*:[104]

> πϱῶτον μὲν οὖν τὰ ὑποτεταγμένα τοῖς φϑόγγοις, ὦ Ηϱόδοτε, δεῖ εἰληφέναι, ὅπως ἂν τὰ δοξαζόμενα ἢ ζητούμενα ἢ ἀποϱούμενα ἔχωμεν εἰς ταῦτα

102. There are numerous other references to "things" in linguistic contexts with Epicurus' fragments. One particularly enticing, if unclear, reference is found in *On Nature* 28 (fr. 13, col. 2 *inf.*—col. 3 *sup.* Sedley = Sedley 1973: 46): ... τὴν τῶν ὀνομάτ[ω]ν κα[ὶ π]ϱαγμάτω[ν ἀ]δι[αλ]ηψία[ν ὁϱ]ῶμ ... ("seeing the confusion of words and things"). What is so striking about this comment from the vantage point of speech acts is that it may indicate that words and things are "confused" precisely because words can be, that is, can perform, deeds.

103. Porphyry *Ad Marc.* 31, p. 209, 23 Nauck = Usener 221, part (Usener 1887: 169).

104. It should be made clear at the outset, although hopefully it will become clear enough in the subsequent discussion, that the present passage and the following one are relevant to the

> ἀναγαγόντες ἐπικρίνειν, καὶ μὴ ἄκριτα πάντα ἡμῖν <ἴῃ>[105] εἰς ἄπειρον ἀποδεικνύουσιν ἢ κενοὺς φθόγγους ἔχωμεν. ἀνάγκη γὰρ τὸ πρῶτον ἐννόημα καθ' ἕκαστον φθόγγον βλέπεσθαι καὶ μηθὲν ἀποδείξεως προσδεῖσθαι, εἴπερ ἕξομεν τὸ ζητούμενον ἢ ἀπορούμενον καὶ δοξαζόμενον ἐφ' ὃ ἀνάξομεν.
>
> ἔ<τ>ι τε κατὰ τὰς αἰσθήσεις δεῖ πάντα τηρεῖν καὶ ἁπλῶς τὰς παρούσας ἐπιβολὰς εἴτε διανοίας εἴθ' ὅτου δήποτε τῶν κριτηρίων, ὁμοίως δὲ καὶ τὰ ὑπάρχοντα πάθη, ὅπως ἂν καὶ τὸ προσμένομενον καὶ τὸ ἄδηλον ἔχωμεν οἷς σημειωσόμεθα. (Epicurus, *Letter to Herodotus* 37–38)[106]
>
> First, Herodotus, we ought to have grasped that which has been made subject to utterances so that we can judge our opinions, queries, and questions by making reference to these things, and that everything not go undistinguished for us, as we make endless demonstrations, or that we make vain utterances; for it is necessary in the first place[107] to regard a concept according to each utterance and to want no additional proof, if indeed we shall have something to which we shall refer our opinion, query, or question.
>
> Further it is necessary to give heed to everything in accordance with our sense perceptions and, simply put, our present focusings, whether of thought or some other yardstick; in like fashion must we also attend to our existing impressions, so that we may have something by which we shall signify both the unclear and what awaits.

Without delving into all of the interpretive quagmires of this passage, we may note that its general structure has at least two clear features: first, there is a temporal movement in the passage from a set of perfect, completed actions (τὰ ὑποτεταγμένα, εἰληφέναι) through to a set of future verbs (ἕξομεν, ἀνάξομεν, σημειωσόμεθα), yet every main verb (whether implied or actual) is in the present

performative-constative tie, precisely because they describe a speech act (nomination) but limit (the conditions of possibility of) that speech act by "context."

105. <ἴῃ> add. Usener | om. BPF. The sigla follow Dorandi (ed.) 2013.

106. The text here, as in general with Diogenes Laertius and the writings of Epicurus preserved there, follows Dorandi (ed.) 2013, although I have compared that text with Marcovich (ed.) 1999.

107. I here agree with Asmis 1984: 31 that there is no need to imply some idea of *prolepsis* in Epicurus' text at this point. In fact, as my translation makes clear, I do not think that τὸ πρῶτον (which I take as adverbial) belongs together with ἐννόημα as a semantic unit. As Diogenes Laertius' gloss makes clear (10.33: παντὶ οὖν ὀνόματι τὸ πρώτως ὑποτεταγμένον ἐναργές ἐστι), the notion of "first" (however we understand it) belongs together with the verbal action in the first sentence of the passage (which of course still allows as many interpretations as we can impute to that "first"). Asmis suggests that even if we take πρῶτον as an adjective, a "first concept" need only be "first" in the sense of what I here call a "given."

tense; second, every main verb or verbal clause expresses obligation or necessity (δεῖ, ἀνάγκη, δεῖ). These general features suggest that the passage is an attempt to state the conditions of possibility of language use in the present time, particularly as regards language about future events. (Such a conclusion would be supported by our earlier encounters with Epicurean dialectic.) The thrust of the passage seems to be that there must be a "given" made subject to utterances (τὰ ὑποτεταγμένα τοῖς φθόγγοις) in order to prevent one of two outcomes: "endless proofs" or "vain utterances" (κενοὶ φθόγγοι). This latter outcome is, of course, familiar from our previous passages, signifying perhaps utterance that has no chance at therapeutic action.

This passage thus makes, in a more positive form, an essentially Derridean point about context. As we have seen, Derrida questions whether there can be a "rigorous, scientific concept" of context, but he also (as he is well aware) relies upon a linguistic context to make this very assertion: this fact is what makes Derrida's notion of context in some ways so paradoxical. "Context" is the necessary buttress for speech, both performative and constative, even if it may never be fully known. Returning to our passage, we note that Epicurus needs a "given," a "context," to speak or "sign" about "the unclear and what awaits," just as the paradigmatic performative, the marriage rite, needs its already established "I do" or *spondeo* (among other things) to occur at all. Put differently, what this passage proclaims is that the performative is in a certain sense predicated upon the (ideal) possibility of the constative: having a "given," a possible (if unrealizable) purely constative function of language, allows us to avoid "endless demonstrations" (καὶ μὴ ἄκριτα πάντα ἡμῖν <ἴη> εἰς ἄπειρον ἀποδεικνύουσιν), that is, endless attempts to create a language with which to make performative (therapeutic) utterances.[108] An initial act of naming (a "demonstration") separate from "context," separate from a prior linguistic "given," would hardly be an act of naming, if by "an act of naming" we mean the speech act of nomination; it would be rather an attempt to create an entire semiotic system, a system which would then allow nomination. While in traditional Greek belief a divine figure like Hermes may accomplish magically the operation of name-giving in a vacuum, our Epicurean sources purposefully stress the impossibility of this act.[109]

108. The precise meaning of the verb ἀποδεικνύειν is debatable. Asmis 1984: 351 defines ἀπόδειξις ("demonstration") as "showing by argument (that is by λογισμός, 'calculation') something nonapparent." This definition sounds a bit too systematic and technical, for λογισμός, however we translate it, is (as *Letter to Herodotus* 75–76 makes clear) the mental faculty employed in coining words. We may, I suppose, call that faculty "calculation," but it is on some level a creative faculty (and not merely an analytical one). Translations such as "reasoning power" (LSJ$^9$ s.v. III) thus seem quite attractive.

109. See, e.g., Lucretius *DRN* 5.1041–45, Diogenes of Oenanda Fr. 10, col. 3, ll. 4–6 Chilton (where Hermes is named).

Such analysis is supported by another famous passage from the *Letter to Herodotus*:

> ἀλλὰ μὴν ὑποληπτέον καὶ τὴν φύσιν πολλὰ καὶ παντοῖα ὑπὸ αὐτῶν τῶν πραγμάτων διδαχθῆναί τε καὶ ἀναγκασθῆναι, τὸν δὲ λογισμὸν τὰ ὑπὸ ταύτης παρεγγυηθέντα ὕστερον ἐπακριβοῦν καὶ προσεξευρίσκειν ἐν μὲν τισὶ θᾶττον, ἐν δὲ τισὶ βραδύτερον καὶ ἐν μὲν τισὶ περιόδοις καὶ χρόνοις †ἀπὸ τῶν ἀπὸ τοῦ ἀπείρου <*>[110], ἐν δὲ τισὶ καὶ ἐλάττους.
>
> ὅθεν καὶ τὰ ὀνόματα ἐξ ἀρχῆς μὴ θέσει γενέσθαι, ἀλλ' αὐτὰς τὰς φύσεις τῶν ἀνθρώπων καθ' ἕκαστα ἔθνη ἴδια πασχούσας πάθη καὶ ἴδια λαμβανούσας φαντάσματα ἰδίως τὸν ἀέρα ἐκπέμπειν στελλόμενον ὑφ' ἑκάστων τῶν παθῶν καὶ τῶν φαντασμάτων, ὡς ἄν ποτε καὶ ἡ παρὰ τοὺς τόπους τῶν ἐθνῶν διαφορὰ ᾖ· ὕστερον δὲ κοινῶς καθ' ἕκαστα ἔθνη τὰ ἴδια τεθῆναι πρὸς τὸ τὰς δηλώσεις ἧττον ἀμφιβόλους γενέσθαι ἀλλήλαις καὶ συντομωτέρως δηλουμένας· τινὰ δὲ καὶ οὐ συνορώμενα πράγματα εἰσφέροντας τοὺς συνειδότας παρεγγυῆσαί τινας φθόγγους, τοὺς <μὲν>[111] ἀναγκασθέντας ἀναφωνῆσαι, τοὺς δὲ τῷ λογισμῷ ἑλομένους κατὰ τὴν πλείστην αἰτίαν οὕτως ἑρμηνεῦσαι. (Epicurus *Letter to Herodotus* 75–76)

Nevertheless, we must grasp that even nature (τὴν φύσιν) has been taught many and varied matters by the things themselves, and forced into things; and reason later rendered more exact the things handed over by nature, and discovered others in addition, among some more quickly, among others more slowly, and in some periods and times increasing more,[112] and in others less.

Whence, too, names did not arise originally by coinage (θέσει), but the very natures of men experienced particular impressions and grasped particular images; and they expelled in a particular way air sent by each of the impressions and images, so that there would be difference among the races across their locales. And later in common, particular additions were coined in each race for making their directions for one another less ambiguous, and more succinctly expressed. Then in introducing some things not generally known, those in the know passed on certain utterances, compelled to proclaim some, choosing

110. Post ἀπείρου lac. indic. Usener: μείζους λαμβάνειν ἐπιδόσεις e.g. suppl. Usener, alia alii.

111. μὲν add. Gassendi

112. I translate Usener's supplement.

> others rationally, they thus interpreted/translated <each term> under the best heading.[113]

This description of the origins of names, that is, the origins of language,[114] is often characterized as, on the one hand, a naturalist response to extreme conventionalism (as advocated by Diodorus Cronus)[115] and, on the other, as an account that marries convention and regional diversity to naturalism.[116] In short, the account proclaims that language is neither exclusively φύσει ("by nature") nor θέσει ("by positing," "by convention"), although it is always, whether in ancient doxographies or in modern commentaries, described as a "naturalist" account of the origins of language. From the standpoint of the present investigation of Epicurean speech acts, what is remarkable is the insistence, as in the previous example, that there is a natural "given" prior to and necessary for any speech act of naming. Thus, names are not something external to "things," applied after "things" come into existence, but rather concomitant with those things, concomitant with a native context, precisely as our previous investigation would suggest.[117] Remarkable, too, is the fact that Epicurean naturalism does not here consist in attributing any specific features (that is, *essential* elements)[118] to the natural context, which precedes any act of naming; rather there is variation among languages based upon the diversity[119] of "things," physical surroundings, experiences, etc., even if that

113. The text here is uncertain: I offer a translation solely as an aid to the reader.

114. It should be emphasized that the plural here ("origins") is intended to indicate how much this passage resists being what it is ("an origin"), positing at once many different origins for language in many different ways at different times.

115. Diodorus Cronus famously named a slave Ἀλλὰ μήν ("however"), apparently "to prove that any word can be used to denote something" (Sedley 1973: 63). For further on Diodorus Cronus, see Gellius *NA* 11.12, with Sedley 1973: 62–65 and 1977.

116. The standard presumption prior to Epicurus seems to have been that language was either a human invention or a god-given gift. See, e.g., Plato *Crat.* 439c2, *Prot.* 322a; Xenophon *Mem.* 4.3.12.

117. As Derrida (1988: 78) remarks, "… contextual difference changes everything, because it determines what it determines *from within*." (Translated by S. Weber.)

118. This fact provides another reason why I am reluctant to understand τὸ πρῶτον at *Ep. Hdt.* 37 as meaning anything other than "in the first place," i.e., as a logical first step to allow further inquiry; other interpretations must grapple with the fact that they give "natural" a much stronger sense than it can bear here. For a similar view, see Schrijvers 1999: 56n7 (originally published 1974).

119. This diversity is emphasized by the repetitious sequence ἴδια … ἴδια … ἰδίως … τὰ ἴδια ("particular … particular … in a particular way … particular additions").

diversity is not so great as to conjure up the notion of a Wittgensteinian private language.[120] (Tellingly, the Hippocratic *Airs, Waters, Places* is often adduced as a *comparandum* to this passage.) Beyond the fact that there naturally is language (it is a fact of nature), there is strikingly (over against earlier naturalist theories)[121] little specificity or content to the naturalness of language.

This passage thus confirms previous analysis—both of the need for context and the variability of that context—but its other aspects may enrich our discussion of the performative-constative tie in Epicurean linguistic thinking. Notably, this passage seems to suggest two types of indistinction at the origins of language: the impossibility of distinguishing finally[122] between, in the first place, actions and things and, in the second, between words and things. These indistinctions are nicely if infuriatingly brought to a head by the final lines of the passage, which are almost impossible to translate: τινὰ δὲ καὶ οὐ συνορώμενα πράγματα εἰσφέροντας τοὺς συνειδότας παρεγγυῆσαί τινας φθόγγους, τοὺς <μὲν> ἀναγκασθέντας ἀναφωνῆσαι, τοὺς δὲ τῷ λογισμῷ ἑλομένους κατὰ τὴν πλείστην αἰτίαν οὕτως ἑρμηνεῦσαι. ("Then in introducing some things not generally known, those in the know passed on certain utterances, compelled to proclaim some, choosing others rationally, they thus interpreted <each term> under the best heading.")

What makes for maddening Greek here is that the utterances (φθόγγους), the "things" (πράγματα), and "the men in the know" (τοὺς συνειδότας) are all in the same grammatical case (the accusative), and it is at times hard to discern who acts on whom, that is, grammatical subject-object relations, or indeed, even more basically, how words should be grouped. The initial portion of the sentence (τινὰ δὲ καὶ ... φθόγγους) seems comprehensible, if a bit overloaded; but its concluding section (τοὺς <μὲν> ... ἑρμηνεῦσαι) is utter chaos, as indicated by the myriad attempts at emendation.[123] The articulation of the sentence suggests that each member of the τοὺς <μὲν> ... τοὺς δὲ ("some ... others") contrast—which is itself partially based upon emendation—should take τινας φθόγγους, the

120. Brunschwig 1977 treats the question of Epicurus and private language, which has sometimes been thought to be a possible interpretation of the ἰδι—words mentioned in the previous note.

121. Earlier "naturalist" theories of language emphasized not the natural origins of language but the natural connection between a name and its referent, that is, the name "naturally" described its referent.

122. I say "finally" because I in no way wish to deny that the distinctions under discussion are useful or impossible to make in a provisional sense.

123. The text printed above accepts (with Marcovich) part of Gassendi's emendation: he originally proposed <καὶ> τοὺς <μὲν>. Sedley (1973:17 and LS) proposes <μὲν οὖν>; Usener deletes τοὺς[1].

immediately preceding object, as its antecedent. The participles within the two clauses (ἀναγκασθέντας, ἑλομένους), if we understand them as agreeing with the immediately preceding articles, seem, however, to argue against this reading: one normally thinks of speakers rather than utterances as "compelled to proclaim" and "choosing by reason." To solve this difficulty, we may, as in the above rendering, separate the contrasting articles from the subsequent participles, a course which provides better sense, if somewhat deceptive syntax.[124]

Whether this interpretation is accepted or not, the difficulty inherent in construing these lines may have a philosophical point; that is, the relative indistinction of grammatical subject and object mirrors the relative indistinction of the act of speaking and the object of speech, of the speaker and the utterance. "Relative indistinction" here does not mean that no distinction may be made, but rather that it is impossible, for example, completely to separate speaker, name, and act in the speech act of nomination. This point is not insignificant, for not only, as Austin reminds us, will nomination misfire if the speaker is a horse (thus demonstrating the inseparability of the speaker from the act), but also an apparently constative description of the conditions of possibility of nomination may itself act upon us, even if primarily in infuriating ways.[125] If Epicurus did, as this discussion has been arguing, uncover a notion of performative speech, then it is worthwhile paying attention not only, as has been chiefly the case thus far, to the constative articulation of the performative but also to the performative aspects of (apparently) constative descriptions.

## (v) *Prolepsis* and the Constative

Any discussion of the constative in Epicurean linguistic thinking, as noted previously, cannot avoid some engagement with the Epicurean technical term *prolepsis*. Here, as a thoroughgoing re-examination of *prolepsis* would take us far from the current topic, I will only outline how *prolepsis* fits with some

---

124. Sedley (1973: 19), who reads <μὲν οὖν> where I read simply <μὲν>, does not think that "those in the know" could possibly be compelled since earlier in the passage it is suggested that the stage of natural compulsion precedes nomination; but I find his reading over-literal. As we have stressed, a natural "given" is necessary for any act of nomination, but such a realization does not mean that context is subsequently fixed, or that nomination is fully calculated.

125. If one employs the terminological pair "performative-constative," then we may say that that which is not purely constative is at least somewhat performative. In one of Austin's other terminologies, we may debate whether the force here described is illocutionary (accomplished in the utterance) or perlocutionary (accomplished through the utterance), although one point of the present discussion has been to cast doubt on whether such a distinction could, in the final analysis, hold up. Benveniste 1966: 270, for one, urges retaining the original "performative-constative" distinction, and most subsequent interpreters have followed suit.

of the other linguistic features described thus far. This discussion therefore makes no claim to being substantially original or complete.

To begin, then, *prolepsis* is perhaps most usefully defined by reference to one portion of the definition provided by Diogenes Laertius (10.33), where he remarks, after a list of technical terms, that a *prolepsis* is "a memory of something having appeared outside several times" (μνήμην τοῦ πολλάκις ἔξωθεν φανέντος). He then offers an example: for the statement "man is such a sort of thing" ('τό τοιοῦτόν ἐστιν ἄνθρωπος'), he suggests that "as soon as 'man' is said, straightaway in accordance with a *prolepsis*, the impression of it is understood, with the senses leading the way." There are two salient features: first, *prolepsis* seems to be the means by which a recognized, general term is interpreted.[126] In other words, as *prolepseis* are empirically formed, they may only exist after repeated sensations of a certain kind of objects, presumably including words themselves; and they therefore—the second point—cannot play an *active* role in the creation of new vocabulary, or even in the creation of new uses of extant vocabulary.

In other words, *prolepseis* would most obviously come into play in attempts at constative language, attempts at describing what is already known. Yet—as we have seen that the constative function of language may never be distinguished fully and absolutely from its performative function—*prolepseis* would inevitably play *some* role (if not an *active* one) in the creation of new words and new meanings, as in performatives more generally. *Prolepseis*, in this sense, would form part of the "given" that must be present in order to allow performative speech. Thus, considering for a moment the speech act of nomination, a statement like "I dub this man Socrates," if felicitous, would be underwritten by the *prolepsis* of man, although there could be no prior *prolepsis* of this uniquely named Socrates. Without the *prolepsis*, the generic "given," the act of nomination not only fails, but it is incomprehensible. Such a function seems simple enough. Yet *prolepseis* are not only a tool for underwriting performative uses of language. As some of our sources make clear, *prolepseis* may also lead to linguistic error.

One of the most important clues for this fact is the following *sententia*:

> ἔνθα μὴ καινῶν[127] γενομένων τῶν περιεστώτων πραγμάτων ἀνεφάνη μὴ ἁρμόττοντα εἰς τὴν πρόληψιν τὰ νομισθέντα δίκαια ἐπ' αὐτῶν

---

126. The evidence is rather unclear, but presumably *prolepseis* would not be necessary for non-generic terms. For a specific individual, for example, I would not need repeated empirical evidence. If I have an adequate memory, presumably meeting Socrates once would suffice for knowledge of his name.

127. καινῶν Aldobrandinus | κενῶν BPFD. The emendations here and in the next note seem necessary to give the text any tolerable sense. If the original readings were retained, however,

τῶν ἔργων, οὐκ ἦν ταῦτα δίκαια. ἔνθα δὲ καινῶν[128] γενομένων τῶν πραγμάτων οὐκέτι συνέφερε τὰ αὐτὰ δίκαια κείμενα, ἐνταῦθα δὴ τότε μὲν ἦν δίκαια, ὅτε συνέφερεν εἰς τὴν πρὸς ἀλλήλους κοινωνίαν τῶν συμπολιτευομένων· ὕστερον δ' οὐκ ἦν ἔτι δίκαια, ὅτε μὴ συνέφερεν. (Epicurus *RS* 38)

Where the circumstances were not new and it was manifest that what was believed just from the facts themselves did not fit our preconception (*prolepsis*), these things were not just. But where the circumstances were new and the same things, those justly disposed, were no longer beneficial, there clearly <these things> were just at one time, since they were beneficial for the community of fellow-citizens towards one another; but later <these things> were no longer just, since they were not beneficial.[129]

Here Epicurus seems to be making a point about context that we have seen several times: if the circumstances, the context, the "given," remain constant, then *prolepseis* (and language based on *prolepseis*) may accurately (and constatively) describe those circumstances. If, however, the circumstances change, then a previous *prolepsis* of justice may be invalid, no longer accurately describing what is just. Such a pronouncement is particularly important if—as we have suggested elsewhere—πράγματα (here "circumstances") include acts, specifically speech acts like nomination.[130] The Epicurean linguistic position will then be a difficult one, for—as we have been suggesting—performative speech, at least in so far as it may be therapeutic speech, is particularly important to the central project of Epicureanism; yet such speech is guaranteed to alter the "circumstances," the "context," and therefore to render previous *prolepseis*, which underwrite not only all general concepts but also the very possiblity of performative

my interpretation of the passage would remain the same, although the second sentence would seem fairly opaque. (Both emendations are generally accepted by editors.)

128. καινῶν Gassendi | καὶ τῶν B, Aldobrandinus.

129. Our evidence about the *prolepsis* of justice is perhaps the best of any surviving evidence on *prolepseis*. In addition to the present *sententia, RS* 37 also discusses the *prolepsis* of justice and *RS* 31–36 all provide useful background information. For further discussion, see Alberti 1995 and Porter 1996: 625–26.

130. To be clear: I have suggested previously that πραγμάτων φθόγγοι ("utterances of things") are speech acts in the sense that we normally use that term (i.e., saying "*spondeo*"); here I mean the accomplished act that results from such utterance (i.e., a marriage agreement *vel sim.*).

speech, insecure.[131] This difficulty may be what Epicurus has in mind when he remarks in the fragmentary remains of his *On Nature*:

> … πᾶσα ἡ ἁμ[α]ρτία ẹ̓στίν | τῶν ἀνθρώπων οὐδὲν ἕτε|ρον ἔχουσα σχῆμα ἢ τọ̀ ἐπὶ | τῶμ πρọλήψεων γιγγ̣[ό]|μενον καὶ τῶμ φαιṿ[ομ]ένων | διὰ τοὺς ̣πολυτρόπους ẹ̓[θι]|σμοὺς τῶν λέξεων … (Epicurus *On Nature* 28 31.10.6–12 = fr. 12, col. 3, ll. 6–12 Sedley)
>
> … all human error has no other structure than that, which arises from *prolepseis* and phenomena due to the much-turning customs of language …

It is not clear from the papyrus remains that Epicurus ever held the position expressed here.[132] But if he did, in light of the discussion conducted to this point, we may surmise that his rationale for holding it found its basis, at least on some level, in his investment in performative speech.[133]

## IV. Conclusions and a Road Map

This review of evidence concerning Epicurus and Epicurean linguistic thinking has provided us with reason to think that Epicureanism manifested an interest in performative speech. The Epicurean tradition preserves two descriptions that accurately depict the relation between signifier and referent in performative speech. Epicurus himself, from his rejection of dialectic to his emphasis on the therapeutic action of language, consistently provides clues that he was interested in thinking about how language acts and not merely about how it describes. Further, we have glimpsed at various points strategies to harness the performative power of language. In particular, our exploration of the evidence from Cicero's *De Fato, Lucullus*, and elsewhere has made clear that it is worth paying attention to Epicurus' engagement

131. Indeed, Epicurus seems to be suggesting that a notion of what is beneficial (συμφέρει) should override any *prolepsis* of justice, although one wonders how one perceives what is beneficial apart from concepts like justice.

132. For further discussion of the textual possibilities, see Sedley 1973: 60.

133. The conceptual role that we have outlined here for *prolepsis* may explain why this term, which was undoubtedly of great epistemological importance in Epicureanism, nonetheless appears so infrequently in extant remains. In the writings of Epicurus himself—his three letters, two collections of aphorisms, and his numerous fragments—the term *prolepsis* is only securely attested five times (a total we may elevate to seven occurrences if we include a pair of speculative emendations). Such a dearth need not indicate a lack of importance; it may indicate rather that Epicurean linguistic thought was heavily oriented toward performative language.

with religion not simply as an antagonistic affair but as an attempt to engage the religious authority that underwrites much performative language.

There is evidence beyond what we have considered, evidence that, while occasionally unclear, may flesh out a portrait of Epicurus' explicit theoretical engagement with performative language. In one anecdotal account, for example, we learn that Epicurus used legal-like formulae to proclaim the topic of discussion (a performative act of naming) but then refused to define that topic (a constative endeavor);[134] in another, we learn that he chided Socratic irony, perhaps a sign that he was sometimes troubled by the misfit between *prolepseis* and present-time (performative) uses of language.[135] Yet even without having the fullest possible picture of Epicurus' linguistic activities, we should feel comfortable in asserting that his investigations share much in common with those of Austin (and others who have taken an interest in performative language). In fact, Shoshana Felman has characterized Austin's inquiries into ordinary language as attempts to uncover the " 'atoms' of language,"[136] so it is perhaps not

134. Cicero *De Fin.* 2.1–2.2: *omnis autem in quaerendo, quae via quadam et ratione habetur, oratio praescribere primum debet ut quibusdam in formulis* EA RES AGETUR, *ut, inter quos disseritur, conveniat quid sit id, de quo disseratur. hoc positum in Phaedro a Platone probavit Epicurus sensitque in omni disputatione id fieri oportere. sed quod proximum fuit non vidit. negat enim definiri rem placere, sine quo fieri interdum non potest, ut inter eos, qui ambigunt, conveniat quid sit id, de quo agatur …* ["Moreover, every discourse that leads an inquiry in a certain fashion and manner ought in the first place to make a *praescriptio* (like "*ea res agetur*" in certain <legal> *formulae*) so that there is agreement between the parties to the discussion as to what it is which is being discussed. This sentiment, which was proposed by Plato in *Phaedrus*, was approved by Epicurus, who felt it appropriate that it (*sc.* a *praescriptio*) occur in every discussion. But he did not see what was next, for he denies that it is appropriate to define the matter at hand, without which definition it is sometimes impossible to be agreed among the parties to the discussion what it is the discussion is about …"]

135. Cicero *Brut.* 292: *tum ille: "ego, inquit, ironiam illam quam in Socrate dicunt fuisse, qua ille in Platonis et Xenophontis et Aeschini libris utitur, facetam et elegentem puto. … decet hoc nescio quo modo illum, nec Epicuro, qui id reprehendit, assentior."* ["Then Brutus: 'I find,' he says, 'that irony, which they say was in Socrates, which he shows in the works of Plato, Xenophon, and Aeschines, witty and elegant.… This fits him in some fashion, and I disagree with Epicurus, who chides this behavior.' "]

136. See Felman 2002: 108: "What Austin analyzes are in a way the 'atoms' of language, that is, 'very little bits of matter' of language, 'that are still just like matter in the fact that they persist through time, and that they travel about in space.' … through the new concept of 'language act' he explodes both the opposition and the separation between matter (or body) and language: matter, like the act, *without being reducible to language*, is no longer entirely *separable* from it, either." (The phrases in single quotation marks are all direct quotes from Austin.) The fact that Austin and Epicurus are both linguistic atomists of a certain stamp should not be taken to mean that they are linguistic reductionists. In fact, both seem to like to identify smaller units of language (performatives, constatives) only to show that those units may not be distinguished fully and absolutely; that, in other words, language is somehow not reducible to the sum of its parts.

so strange that his pursuits fit well with those of an ancient atomist.[137] One should not, of course, overlook differences: Epicurus' ethical interests (e.g., his psychological hedonism), for example, do not find obvious parallel in Austin's writings (although Austin was a professor of moral philosophy). The goal of this discussion is not to turn Epicurus into a twentieth-century English philosopher so much as to give weight to a body of evidence, particularly evidence concerning Epicurus' rejection of dialectic, which is often ignored in trying to reconstruct his epistemology and linguistic thinking.

In particular, the present discussion differs from previous accounts in its emphasis on context, which is, as earlier theoretical discussions made clear, a natural consequence of its focus upon performative language. Indeed, as we move into the text of Lucretius, cultural and historical evidence, which allows reconstruction of rites, rituals, and performative language, will play an increasingly significant role. In subsequent chapters, I first investigate two different speech acts, promising in one chapter and nomination (specifically the use of proper names) in the next, both of which play an important role in the overall (political and therapeutic) project of the *De rerum natura*. In conclusion, the discussion returns to origins—that is, the origins of language and human society—in an attempt to gain a better grasp of the political implications of Lucretian speech acts. The argument proceeds, then, not by applying a modern theory to Epicureanism or *De rerum natura*; but rather by tracing the tracks of Epicurean linguistic thinking itself, the investigation leads to new conclusions about what it means to place Lucretius into context, to locate his "outside."[138] As the study of speech acts teaches, it is only by attending closely to the text, by attending closely to the "inside" and not—as Lucretius sometimes suggests of his addressee Memmius—drawing back from it, that we may find that "outside."[139]

---

137. Derrida, too, sees connections between his work and Epicureanism, although he is more interested in the *clinamen* (the atomic "swerve") than in atomism per se. See Derrida 1984.

138. It is of course the case, as has been explicitly true in the present introduction, that modern theories (whether we like it or not) infect our reading and reconstruction of Epicurean thought; but the point is that any ahistoricism caused by such theories is always already there. It is nevertheless important to point out that the argument of this introduction (and this work as a whole) works on two—not wholly distinct—levels: it first aims to motivate a certain reading of Epicurean linguistic thought, and then to use that reading to read an Epicurean text, the *De rerum natura*. The result of this investigation is, however, intended to answer modern problems, at least in the sense that it aims to contribute to a discussion of "inside" and "outside," of performance and therapy.

139. See, e.g., *DRN* 1.410: *quod si pigraris paulumve recesseris ab re* . . . ("But if you drag your feet and draw back a bit from my subject . . .").

# 2

# *Lucretian Promises: Fides, Foedus, and the Politics of Nature*

*Ein Tier heranzüchten, das versprechen darf—ist das nicht gerade jene paradoxe Aufgabe selbst, welche sich die Natur in Hinsicht auf den Menschen gestellt hat? ist es nicht das eigentliche Problem vom Menschen?*[1]

*... two orders that, in reality, do not communicate: the order of the act and the order of meaning ...*[2]

IF SPEECH ACT theory is native to Epicurus' philosophy—that is, if Epicureanism defines performative language even in some of its most technical details—this observation would not necessarily make performativity a central, thematic concern of Lucretius' *De rerum natura*. This chapter establishes more fully how this is so. It tracks Lucretius' engagement with a specific type of performative utterance, the promise, and more generally with terminology surrounding promises, in particular, the terms *fides* ("trust, trustworthiness") and *foedus* ("compact, agreement"). The promise, or at least ideas that one may extract from it, plays a central role in linguistically figuring two questions treated by Lucretius' text: first, the question of how to delimit material boundaries, particularly the boundaries of our sensory encounter with the world; second, the question of how to assign agency and understand action in the material world. Before turning to the details of promises and promising,

1. Nietzsche 1988: 5.291 (*Zur Genealogie der Moral* 2.1): "To breed an animal that *may promise*—is that not precisely the very paradoxical task that Nature set herself with respect to humankind? is that not the true problem *of* humans?"

2. Felman 2002: 17.

however, let us—through an argument linking performative language with power (*posse*)—motivate and contextualize the study of performatives within Lucretius' text more generally.

## *I. Performatives and Power*

Lucretian studies have seen relatively little discussion of speech act theory to date.[3] Yet Lucretius' contemporary Catullus has been fertile ground for the investigation of performative language. As Daniel L. Selden remarks in a seminal essay ("*Ceveat lector*: Catullus and the Rhetoric of Performance") for the use of speech act theory in studying Latin poetry,[4]

> ... the majority of Catullus' poetry is in fact performative and not constative.... Whereas the principal burden of his contemporaries' compositions—Lucretius, Cicero, Varro Atacinus—tends to be descriptive, that is to record *res gestae* or explain the phenomena of nature, Catullus' texts are for the most part incidental and almost always have a specific act in view.... the collection as a whole reads virtually as a catalogue of different discursive operations: dedication (*c.* 1), censure (*c.* 39), advice (*c.* 69), condolence (*c.* 96), warning (*c.* 21) ... and so forth.[5]

---

3. There has been little explicit discussion of speech act theory within Lucretian scholarship, although there has been quite a bit of discussion that articulates claims (explicit or implicit) of speech act theory. Kennedy 2002 and Gale 2004b, for example, make points (Kennedy on constructivism, Gale on the *poeta creator*) that may be subsumed under a theory of how the poet/speaking agent constructs and creates rather than merely describing. Edmunds 2001: 27–28, in a broader consideration of the application of speech act theory to literature, denies that Lucretius' poem itself is a speech act, reserving that term for utterances performed in ritual contexts. Lowrie 2009: 32–34 offers a brief discussion both of Lucretius' representation of his own poetic media and his poem's pragmatic aims. Marxist readings of Lucretius (which have a long history, including—perhaps most famously in English—Farrington 1965) also have sympathies with speech act theory, but there are important differences, particularly with regard to Marx's view of history. A close point of comparison, noted at Selden 1992: 481 (Selden 2007: 522), is Marx's eleventh *These über Feuerbach*: "Die Philosophen haben die Welt nur verschieden interpretiert; es kommt aber darauf an, sie zu verändern."

4. There are many other studies within the larger discipline of Classics that draw upon speech act theory (in various guises). One thinks, for example, of the work of Gregory Nagy (e.g., Nagy 1990) and R. P. Martin (Martin 1989) as well as many different pieces by John Henderson (e.g., Henderson 1997 on "naming"). More recent are Lowrie 2009 and Lowrie 2010.

5. Selden 1992: 481–82 (Selden 2007: 522–23).

In many respects, this characterization points to a noteworthy difference between Catullus and Lucretius: Catullus' poetry, as Selden suggests, is "for the most part incidental" while Lucretius' poem is lengthy[6] and sometimes portrayed as merely describing "the way things are."[7] Yet, as we shall see, Selden's own essay provides a model for realizing that Lucretius' poem is not, nor should it be considered, simply descriptive.[8]

It is easy enough to point to individual passages where Lucretius' text manifests, or may be thought to manifest, a performative character.[9] Casting only a quick glance at a few lines from the opening of Lucretius' poem, the famous invocation of Venus Genetrix, we note multiple standard performatives:

> *quae quoniam rerum naturam sola gubernas,*
> *nec sine te quicquam dias in luminis oras*
> *exoritur neque fit laetum neque amabile quicquam,*
> *te sociam* ***studeo*** *scribendis versibus esse*
> *quos ego de rerum natura pangere* ***conor*** *. . . (DRN 1.21–25)*
>
> *You (sc. Venus), since you alone govern the nature of things,*
> *and without you neither does anything arise into shining*
> *shores of light,*
> *nor does anything come to be fertile or capable of love;*

6. The massive size of Lucretius' text—in length, his books are more reminiscent of Apollonius of Rhodes than Virgil or Ovid—plays a certain role in the debates about whether his poem is epic, didactic epic, or didactic *tout court*. These generic issues, while they will no doubt creep into the present discussion at times, are largely sidestepped here, as a pragmatic speech category (the promise) guides the discussion. On genre in Lucretius, see Mayer 1990 (on Lucretius as epic), Gale 1994 (especially Ch. 3, following Mayer), Volk 2002a (especially its introduction, which formulates a definition of didactic as its organizing principle), and Volk 2002b (a review of Gale 2001a, which contests the usefulness of the term "didactic epic").

7. This is Rolfe Humphries' rendering (Humphries 1968) of Lucretius' title.

8. There are many ways of reading performatives in texts. Almost any statement in a text, even the most banal, apparently descriptive utterance may be considered performative if we suppose an author utters "I declare . . ." behind his text. While "The grass is green" appears descriptive, "I declare the grass to be green" appears performative. The present discussion is less concerned with these implicit performatives than with more obvious performatives, often located in ritual contexts. Some performatives considered here will be implict, though, at least in the sense that the actual declarative language of the performative will be omitted.

9. I accept (a premise that some philosophers, though few students of literature, would deny) that one may usefully study speech act theory in conjunction with "literary" works. It is debatable whether Lucretius would be subject to the common critique of literary speech acts—that is, that they are only "pretend"—since the pragmatic, didactic intent of his work is rarely, if

> *you* ***I seek*** *as an ally for writing my verse,*
> *which* ***I try*** *to compose on the nature of things . . .*

Various features of these lines merit comment, but for present purposes it is sufficient to note that, in a rather uncannily classical fashion, the poet here presents two performatives, specifically, two commissives, that is, performatives that "commit the speaker to a certain course of action."[10] With the two verbs *studeo* and *conor*, as is often the case in prayers and invocations, Lucretius lays out his intentions, his wishes, rather than (as he does earlier) chiefly describing Venus or her powers and effects upon the world. Moreover, although many performatives do not meet this grammatical criterion, these two verbs are expressed in the first person, as are Austin's most famous, canonical examples (e.g., "I do" [in a marriage ceremony]).[11]

That Lucretius sometimes employs performative language is hardly surprising, particularly in a hymnic scene where he calls the goddess Venus to his aid. Indeed, such an observation would not alone motivate an extensive study of performatives in *De rerum natura*. Yet Selden's reading of a central poem in the Catullan corpus provides better basis for more extensive inquiry. Though sometimes ignored due to its obscene content, *c.* 16, Selden contends, is programmatic for Catullus' work.[12] He reads this poem against contemporary rhetorical teaching in a fashion that is instructive for the study of performative language more generally. His interpretation contrasts performative language as one, if not the only, language of power with constative utterance as a language of being, that is, a language of how things are:[13]

> *pedicabo ego vos et irrumabo,*
> *Aureli pathice et cinaede Furi,*
> *qui me ex versiculis meis putastis,*

ever, questioned. For defenses of the relevance of speech act theory to literature, see (in addition to Derrida 1988, discussed in the first chapter) Felman 2002 and Johnson 1980: 52–66.

10. Austin 1975: 157.

11. See Austin 1975: 56 on this grammatical criterion (as well as its ultimate insufficiency for determining a performative utterance).

12. The poem is omitted, for example, from C. J. Fordyce's commentary (Fordyce 1961).

13. As has been mentioned in earlier discussion, it is perhaps most accurate to say that "performative" and "constative" describe two idealized functions of language rather than two separate types of utterance. As Austin himself remarks (Austin 1975: 142), "[c]an we be sure that stating truly is a different *class* of assessment from arguing soundly, advising well, judging fairly, and blaming justifiably?" See, too, Austin 1975: 67 (the beginning of the sixth lecture): ". . . the performative is not altogether so obviously distinct from the constative . . ."

*quod sunt molliculi, parum pudicum.*
*nam castum* **esse** *decet pium poetam*
*ipsum, versiculos nihil necesse* **est;**
*qui tum denique habent salem ac leporem,*
*si sunt molliculi ac parum pudici,*
*et quod pruriat incitare* **possunt,**
*non dico pueris, sed his pilosis*
*qui duros nequeunt movere lumbos.*
*vos, qui*[14] *milia multa basiorum*
*legistis, male me marem putastis?*
*pedicabo ego vos et irrumabo. (Catullus c.16)*

*I'll fuck you in the face and the ass,*
*Aurelius, you pathic, and Furius, you fag,*
*you, who from my verses—because they're a bit soft—*
*thought me not pure enough;*
*for it's fitting that a poet* **is** *chaste and dutiful*
*himself, but his light verse need not* **be** *so at all.*
*Verse indeed is charming and witty,*
*if it is a bit soft and not too pure,*
*and* **can** *incite an itch—I don't mean*
*in boys but in these hairy men,*
*who can't move their hard loins.*
*You, who've read many thousands*
*of kisses, thought me not manly enough?*
*I'll fuck you in the face and the ass.*

In his analysis, Selden points to the paradoxical character of the defense Catullus asserts against Aurelius and Furius: "if the poet is actually virtuous and chaste, he will never carry out the rape, and, if he carries out the rape, he substantiates the claims against his [*sc.* own] morals." In other words, with its paradox or its logical contradiction, "[t]he piece not only warns its readers off of any access to the writer through his text, but is specifically set up to block that passage."[15] Selden further sets the paradoxical nature of the poem against traditional rhetorical procedure as

14. vos qui *scripsit nescioquis aevo medio* (= δ *in editione Mynors*) | hosque *O* | vosque *G* | vos quod *Mynors*. Apart from this reading, the text follows the Oxford Classical Text of R. A. B. Mynors (Mynors 1958); the sigla are also Mynors'.

15. Selden 1992: 478 (Selden 2007: 517).

embodied by the triad *docere-delectare-movere* ("to inform or teach," "to delight," "to move").[16] The poet teaches that he is chaste (*castum esse*); he delights with his verse (*salem ac leporem*); and that verse can move to sexual excitement (*quod pruriat incitare possunt*). "While each of these functions [*sc. docere-delectare-movere*] may make separate demands and entails its own agenda, the skillful speaker is conscientiously at pains to minimize their difference."[17] But Catullus employs this trusted triad improperly: he "mischievously parades before his audience the inauthenticity of his own claims."[18] "As a whole, then, *c.* 16 is a puckish travesty of orthodox rhetorical prescriptions. The poem not only cuts the knot that binds oratorical instruction (*docere*) to arousal (*movere*); it develops their capacities to unexpected ends, without in the least diminishing the composition's charm (*delectare*)."[19] Selden sums up his reading as follows:

> Within the framework of rhetorical teaching at Rome ..., Catullus' insistence that we distinguish the information from the outcome of his work also constitutes a serious challenge to the dominant critical position. "It is one thing (*aliud*)," Cicero concedes, "to set forth empirical events ..., but another (*aliud*) to stir up the hearer,"[20] and *c.* 16 purports to press the nature of that difference. By denying authorial indecency at the same time that it claims to incite illicit desire, the poem plays off the referential value of its statements (*esse*) against their predictive power (*posse*).[21]

The opposition between *esse* (constative) and *posse* (performative) describes a distinction that may be generalized beyond *c.* 16. The idea that what a poem means and what it does may not coincide haunts Lucretius' work, with all its epic size, no less than the "puckish" proclamations of Catullus.

Writing at roughly the same time and against the same cultural background as Catullus, Lucretius manifests a similar interest in the juxtaposition of power (*posse*) and being (*esse*).[22] Our examination of Epicurus' letters, among other

16. This triad, which describes ideal rhetorical performance, appears often in Cicero's rhetorical corpus, see, e.g., *De oratore* 2.115, *Brutus* 185, *Orator* 69.

17. Selden 1992: 479 (Selden 2007: 518).

18. Selden 1992: 479 (Selden 2007: 518).

19. Selden 1992: 480 (Selden 2007: 520).

20. Cicero *Opt. Gen.* 15.

21. Selden 1992: 480 (Selden 2007: 520).

22. On the influence of rhetoric (a key background element for Catullus) upon Lucretius (particularly with regard to its import for his argumentative structure), see Asmis 1983, who

evidence in the first chapter of this study, already has shown that his school had a particular interest in language about the future. To name only two salient features, Epicureanism's consideration of psychic therapy and death shows, among other things, that it addresses the possible (e.g., freedom from pain) and the impossible (e.g., sentient experience of death). It addresses, in other words, constraints on how the future may turn out. A similar focus on power and its limits appears throughout Lucretius' poem, beginning in fact with its title. From a purely lexical standpoint, *De rerum natura* may be rendered, albeit tendentiously, as "On Possible Contingencies" (*OLD*[2] s.v. *natura* 6b).[23] Even if this rendering misses the mark (with its overemphasis on possibility and chance), Diskin Clay has brought out how strongly the poem presents *rerum natura* as a "birth of things" in its opening lines.[24] *De rerum natura*, in other words, is clearly concerned with the generative power (*posse*) of the natural world. The fruit of Epicurus' conquest of traditional religion is, as Lucretius describes it, the knowledge of "what can arise,/ what cannot, in which way, finally, power is defined/ for each and its deep-sticking boundary stone" (*DRN* 1.75–77: . . . *quid* ***possit*** *oriri,/ quid* ***nequeat****, finita* ***potestas*** *denique cuique,/ quanam sit ratione atque alte terminus haerens*). This description crams three words for power into two lines, and their sentiment echoes throughout the poem, recurring in more or less the same form at *DRN* 1.594–96, 5.88–90, and 6.64–66. These observations make it hardly surprising that *posse* is one of the most common words in the *De rerum natura*, appearing more than five hundred times in the poem. If we include *queo* ("I am able") and *nequeo* ("I am unable") in our count, the tally rises above 650, making power arguably the most prominent verbal, if not nominal or conjunctive,[25] idea in the entire poem.[26]

Such evidence, which could be multiplied and examined in greater detail, must be contextualized to show its full significance. In Selden's reading of *c.*16, he takes Catullus as offering, at least in part, a constative description

---

cites earlier literature on the subject, and Marcovic 2008. I follow Selden in rendering *posse* with "power," although one may translate it with "potentiality," among other renderings.

23. As the passage at *DRN* 1.21–25 (cited *supra*) makes clear, Lucretius effectively identifies the power that is a central topic of his poem with the (generative, constructive) power of Venus. There is debate about the meaning of Lucretius' title: most link it to the pre-Socratic tradition of works written περὶ φύσεως, although there is disagreement about how directly *natura* renders φύσις. Lévy (ed.) 1996: 18–19, for example, denies that *natura* can mean birth.

24. Clay 1998: 125.

25. The conjunctions *-que* (occurring 1,283 times) and *et* (1,227) as well as the noun *res* (628) may be considered, at least numerically, more prominent than *posse*. See the following note for further details.

26. The numerical evidence is as follows: *posse* occurs 517 times (eighth most common word in the poem); *queo* occurs 84; *nequeo* occurs 56. As a verb, *posse* is only surpassed in sheer number

of his poetic project, a description that portrays a disjunct between what his poem may do and what it says (both about him as its author and about his relationship toward Aurelius and Furius). We may question whether it is safe to presume that Catullus' description is fully constative, but Selden's analysis helps give significance to the following lines of the *De rerum natura*:[27]

> *denique iam quoniam generatim reddita finis*
> *crescendi rebus* ***constat*** *vitamque tenendi,*
> *et quid quaeque* ***queant*** *per foedera naturai,*
> *quid porro* ***nequeant****, sancitum quandoquidem exstat,*
> *nec commutatur quicquam, quin omnia* ***constant***
> *usque adeo, variae volucres ut in ordine cunctae*
> *ostendant maculas generalis corpore inesse,*
> *immutabili' materiae quoque corpus habere*
> *debent nimirum. nam si primordia rerum*
> *commutari aliqua* ***possent*** *ratione revicta,*
> *incertum quoque iam* ***constet*** *quid* ***possit*** *oriri,*
> *quid* ***nequeat****, finita* ***potestas*** *denique cuique*
> *quanam sit ratione atque alte terminus haerens,*
> *nec totiens* ***possent*** *generatim saecla referre*
> *naturam mores victum motusque parentum. (DRN 1.584–98)*

> *Further, then, since a limit of growth has been*
> *given to things by their genus, and a limit to life* ***established;***
> *and seeing that it stands ratified through the compacts of nature,*
> *what each* ***is able****, and what each* ***cannot,***
> *nothing is changed. Instead, all things are* ***established***
> *to such a degree that diverse birds, all of them in order,*
> *show that the marks of the genus are present upon the body.*

---

of occurrences by the linking (not the existential) verb *sum*, which appears 1,330 times. By comparison, *natura* ("nature"), which is by any measure a significant idea in the poem, occurs 236 times; *corpus* ("body, atom"), 542. All word counts here follow the published concordance of Govaerts 1986, which takes the 1972 publication of Alfred Ernout's Budé edition as its base text. Different editions would no doubt produce somewhat different results, but the relative preponderance of terms would remain much the same. Numerical tallies only provide a vague sense of the significance of these various terms for the poem; but insofar as power may be linked with performative language, such counting confirms that performative language plays a substantial role in the *DRN*.

27. Do Catullus' lines simply describe the "fact" that there is no connection between the poet's life and his verse? Or is Catullus trying to persuade us of this claim?

*It is no wonder, too, that they are bound to have a body*
*of changeless matter; for if the first-beginnings of things*
***were able** to be changed, conquered by some rationale,*
*it would then **render** uncertain what **can** arise,*
*what **cannot**, in which way, then, **power** and*
*its deep-sticking boundary are defined for each;*
*and generations **would** not **be able** so many times, genus by genus,*
*to bear the nature, customs, life, and movements of their parents.*

This passage, which occurs within a series of arguments exploring the interrelation of atom and void, emphasizes the perceived constancy of the natural, biological world. In establishing a parallel between the constancy of the macroscopic world of birds and other *genera* and the microscopic world of atom and void, Lucretius proclaims his interest (and investment) in what one might call the constative of the performative, that is, in describing verbally the laws of what can and cannot arise.[28] Thus, just as in *c.* 16 Catullus describes the "laws" of interpreting his verse, explaining what conclusions may be drawn about him as author (none) and what may arise (sexual excitement) from reading that verse, so here Lucretius describes the natural world, articulating the limits of its possibilities.

Lucretius' subject in this passage is, of course, the material world, the world of atoms and void, and not performative language per se. Yet there are at least two senses in which performative language is directly relevant to the present context. First, any full description of what can arise and what cannot must include a description of how language itself arises and how that language creates performatively. Indeed, recognizing such a need, Lucretius provides an account of the origins of language (and its creative power) in his fifth book, a passage considered in some detail in the fourth chapter of the present study. More immediately, however, we may interrogate whether Lucretius here is not so much describing

28. "Performative" here includes all that may arise in (and from) Lucretius' description of atoms and void and thus remains linguistic, if only partially so. The next paragraph makes clearer the specific role that performative language plays in this account. If "performative" here seems to have increased its semantic range, it should nonetheless be noted that insofar as performative language is always embedded in some kind of ritual or regularized activity, "performative" may describe quite a range of behaviors, including literally non-verbal ones. J. Hillis Miller is perhaps the reader who has stretched performative language to its greatest extremes in reading kisses and "sign acts" under the theory of performative language. See Miller 2005: 30–32, who traces his "sign acts" back to Austin's comments on the performative effect of "cocking a snook," throwing tomatoes, and similar acts.

the already extant "compacts of nature" (*foedera naturai*) as creating them, or trying to create them, with his performative poetry. The point is not that Lucretius, in god-like fashion, directly and immediately controls atoms and void with his language (although there is little doubt that atoms and void would be involved in any act of writing or speech),[29] but rather, as he himself emphasizes, that the linguistic figuration of external reality matters for the understanding of that reality, and that he is constantly, performatively creating a new philosophical and scientific vocabulary.[30] Whereas Catullus maintains the *persona* of a poet who dictates what his poetry may and may not do, Lucretius takes an omniscient, god-like perspective, describing (*esse*) all that may arise (*posse*).

Lucretius perhaps most famously thematizes his acts of linguistic creation at *DRN* 1.136–39 (*nec me fallit Graiorum obscura reperta/ difficile inlustrare Latinis versibus esse,/ multa novis verbis praesertim cum sit agendum/ propter egestatem linguae et rerum novitatem* ... "And it does not escape me that it is difficult to cast light upon the dark discoveries of the Greeks in Latin verse, especially since many things must be presented with new words due to the poverty of our tongue and the newness of things ..."), where the term *reperta* presents a central ambiguity for the interpreter. As usually understood, *reperta* are the "discoveries" of Epicurus, but the term may just as well designate (cf. *OLD*[2] s.v. *reperio* 6) Epicurus' "inventions."[31] The point in rendering *reperta* as "inventions" is not that Epicurus created the natural world *tout court* but that the natural world does not exist (certainly not for Lucretius as poet, perhaps not ever) apart from representation.[32] The problem confronting Lucretius—no less than Epicurus, *mutatis mutandis*—may be something greater than the mere "poverty" of Latin. Latin, one may suggest, is not a stable entity used only constatively but also a performative one that itself changes (through acts of naming and catachreses) in Lucretius' hands. There are, in fact, a number of passages throughout the *De rerum natura* where the text proclaims that the poet (or a combination of the reader and the poet) is involved in an act of creation rather than one of description.

One such set of passages is marked by the presence of what Godo Lieberg has called the *poeta creator* ("poet as creator").[33] These passages give the sense

29. On speech in Lucretius, see *DRN* 4.549–52, with Holmes 2005a: 534–46.

30. In this sense the passage would no longer, as in Selden's reading of Catullus, provide a constative of the performative but rather a simple performative establishing the possibility of the constative.

31. See Kennedy 2002: 17–21, contrasting the languages of "discovery" and "invention."

32. See Schrijvers 1970: 63–4 and Waszink 1949 (noted at Long 2006: 214).

33. See Lieberg 1982: 112–14 on Lucretius. Note, too, Hardie 1986: 173 and especially Gale 2004b: 59.

that we (that is, poet and reader, or narrator and narratee) somehow matter in the unfolding (that is, the creating) of the *rerum natura*. At *DRN* 5.110–21, for example, Lucretius writes:

> *qua prius aggrediar quam de re fundere fata*
> *sanctius et multo certa ratione magis quam*
> *Pythia quae tripode a Phoebi lauroque profatur,*
> *multa tibi expediam doctis solacia dictis;*
> *religione refrenatus ne forte rearis*
> *terras et solem et caelum, mare sidera lunam,*
> *corpore divino debere aeterna manere,*
> *propterea que putes ritu par esse Gigantum*
> *pendere eos poenas immani pro scelere omnis*
> ***qui ratione sua disturbent moenia mundi***
> ***praeclarumque velint caeli restinguere solem***
> ***immortalia mortali sermone notantes . . .***
>
> *On which point, before I shall come to pour forth my proclamations*
> *in fashion more holy and with reason much more sure*
> *than that which the Pythia proclaims from the tripod of Apollo*
> *and the laurel, many comforts I shall put forth to you with learned speech;*
> *lest by chance you think, restrained by religion,*
> *that earth and sun and sky, sea stars moon,*
> *ought to remain eternally with divine body,*
> *and lest therefore you consider that, in the fashion of the Gigantes,*
> *all those pay for their gross misdeed, who*
> ***shake the walls of the world with their reason and***
> ***who desire to quench the shining sun of heaven,***
> ***marking things immortal with their mortal speech . . .***

In a passage that may stand as emblematic for the motif of the *poeta creator*, Lucretius intrudes upon his own narrative. The poet speaks of "marking" (*notare*) "immortal" things with mortal speech at the same time as he himself marks those very immortal things (that is, celestial bodies) with his own speech.[34] In other words, he not only thematizes (that is, describes) proclaiming the mortality of the world, but he also proclaims—in a fashion "more holy" (*sanctius*)

34. For further *poeta creator* passages, see, e.g., *DRN* 1.381–82, 2.560–62, 3.14–17. The fact that the passage is in the future tense is not irrelevant to its "prophetic" character. Compare *fabor*

than, if not entirely unlike, the prophetic Pythia—that mortality himself. He performs and describes. Within the space of a few lines, his text is at once constative and performative. The performative aspect of these lines may be infelicitous, and its constative aspect may be false;[35] but we begin to glimpse that an investigation into performative language within the *De rerum natura* may bear fruit, and that its reach extends beyond the initial invocation of Venus Genetrix.

The occurrence within the quoted passage of multiple words derived from *fari*—*fata* (110: "proclamations") and *profatur* (113: "proclaims")—is hardly accidental. As we have noted, *fari* is a special term in the Latin lexicon. Paradigmatically, *fari* is "an absolute assertion that cannot be contradicted because it is not limited to the act of speaking: it simultaneously actualizes the utterance."[36] Modeled on divine speech,[37] *fari* and its congeners[38] are often ($OLD^2$ s.v. *for* 2c[39]) used to name prophetic utterance (speech whose efficacy is guaranteed by the *auctoritas* of a divine principal), as in the following lines from Lucretius' forebear, Ennius (*Ann.* 1.13.15–16 Skutsch): … *Anchisesque Venus quem pulcra dearum/ fari donavit, divinum pectus habere* … ("… Anchises, to whom Venus, beauty of goddesses, granted *fari*, that is, to have a divine [i.e., prophetic][40] chest …").[41] In the Lucretius passage, we may surmise that the density of terms related to *fari* forms part of the poet's engagement with what Philip Hardie calls the "*vates*-concept."[42] Although, as Hardie and others have pointed out, Lucretius (not unlike Ennius before

(which does not occur in the present tense, first-person form **for*), together with the remarks of Bettini 2008: 313–14, 370–71. Homer is not dissimilar: Christensen 2010.

35. Lucretius, that is, conceivably may be a false prophet, both in terms of his message and in terms of the act of prophesizing.

36. Bettini 2008: 315.

37. Pötscher 1978b: 394–409 makes clear the religious basis of *fatum* ("fate"). See, too, Isid. *Orig.* 8.11.90: *fatum autem dicunt esse quidquid dii fantur, quidquid Iuppiter fatur.* ("They say, moreover that *fatum* is whatever the gods say [*fantur*], whatever Jupiter says [*fatur*].")

38. See, e.g., *profor* ($OLD^2$ s.v. 2).

39. Compare the *TLL* (6.1.5.1028–32) article of Vollmer, where he writes (1029): *vis prisca verbi elucet inde quod hic illic significat "loqui cum gravitate" "prophetari"* … ("The ancient force of the term shows through when, here and there, it means "to speak with gravity" "to prophesy' …"). But, as Bettini 2008 (whose evidence ranges over the whole of Latin antiquity) and Pötscher 1978b (whose evidence is primarily imperial) convincingly show, "to prophesy" is not merely an ancient (or poetic) meaning of *fari*.

40. On rendering *divinus* by "prophetic," see Skutsch 1985: 174, with his parallels.

41. For Lucretius' relation to Ennius, see *DRN* 1.112–26.

42. On "Lucretius and the *vates*-concept," see Hardie 1986: 17–22. See, too, Conte 1994a: 4–19. *vates* presents difficulties for the translator but means something like "prophet-seer-poet."

him[43]) disparages the "beliefs and threats of *vates*" (*DRN* 1.109), he nonetheless, drawing inspiration from Empedocles,[44] "adopts for himself . . . [the] stance of the prophet-poet."[45] Particularly striking in this connection is Ovid's comment (*Tristia* 2.425–26): . . . *explicat ut causas rapidi Lucretius ignis,/ casurumque triplex* ***vaticinatur*** *opus* . . . (". . . as Lucretius unfolds the causes of greedy fire, he **prophesies** that the tripartite work will perish . . . ").[46]

The association of Lucretius' poetic voice with a religious function and more specifically with performative utterance is furthered by a passage that immediately precedes the one cited previously:

*nec me animi fallit quam res nova miraque menti*
*accidat exitium caeli terraeque futurum,*
*et quam difficile id mihi sit pervincere dictis;*
*ut fit ubi insolitam rem adportes auribus ante,*
*nec tamen hanc possis oculorum subdere visu*
*nec iacere indu manus, via qua munita fidei*
*proxima fert humanum in pectus* ***templa****que mentis.*
*sed tamen* ***effabor****. dictis dabit ipsa fidem res*
*forsitan, et graviter terrarum motibus ortis*
*omnia conquassari in parvo tempore cernes.*
*quod procul a nobis flectat fortuna gubernans,*
*et ratio potius quam res persuadeat ipsa*
*succidere horrisono posse omnia victa fragore. (DRN 5.97–109)*

*And I am not unaware how the future destruction of heaven and earth*
*strikes the mind as a new and wondrous thing; and*

---

Beyond Hardie, see Newman 1967 (on the *vates* in Augustan poetry), which has received much scholarly discussion.

43. For Ennius on *vates*, see *Annales* 13.4.374 Skutsch, where the editor asserts (1985: 540): "the word here is . . . disparaging in sense."

44. Conte 1994a, Sedley 1998, Sedley 2003, and Garani 2007 (all anticipated by Jobst 1907) collectively make a convincing case for Empedocles' influence on Lucretius. Sedley's clear separation (in Sedley 1998) of "content" and "form," though, simplifies the reality. Study of performative language shows that issues of "language" and "style" are not so separate from more traditionally philosophical issues of doctrine and argument, at least insofar as these latter are aimed at persuasion or psychic transformation.

45. Hardie 1986: 19. Epicurean philosophy, too, styled itself as prophecy.

46. In presenting Lucretius as a *vates*, Hardie (1986: 20) points especially to *DRN* 1.922–32, where the poet speaks of his own inspiration.

*how difficult it is for me, with my words, to convince you thoroughly of this.*
*So it happens when you bring a previously unfamiliar matter to the ears—*
*and yet you could not submit it to the viewing*
*of your eyes, nor toss it in your hands,*
*in which way the paved path of trust leads most directly to the human heart*
*and the* ***sacred precincts*** *of the mind. Nevertheless,* ***I shall proclaim.***
*Perhaps the thing itself will grant belief to my words, and*
*in a moment of time you will perceive everything to be shaken up together*
*by heavy movements of the earth.*
*May governing fortune steer such destruction far from us,*
*and may reason rather than the thing itself convince that all can*
*fall, conquered by fearful-sounding destruction.*

In addressing a topic, the possible destruction of the world, that he senses will be difficult for his audience to accept, Lucretius unmistakably evokes a well-attested Roman ritual, the establishment of sacred space by the *augures* ("priests of the augural college"). As Varro remarks (*LL* 6.53), ... *effari templa dicuntur: ab auguribus effantur qui in his fines sunt* ("sacred precincts are said to be proclaimed: the boundaries for them [i.e., the *templa*] are proclaimed by the *augures*"). Although the specific details of this ritual are, as often, a bit obscure, the collocation of *effari* and *templa* (in Lucretius' as in Varro's text) is repeated in the numerous references to it: Cicero, for example, remarks (*Leg.* 2.21) of the *augures*: ... ***templa*** *liberata et* ***effata*** *habento* ("... may they keep **the sacred precincts** 'released' and '**proclaimed**' ").[47] Thus, Lucretius' text, by evoking such recognizable religious language, paints the poet himself as an *augur*, who with his discourse attempts to reconfigure and "liberate" the *templa mentis* ("sacred precincts of the mind"), to the end of convincing the hearer of his claims about the world's future destruction.[48]

One could linger, analyzing this passage further or collecting more extensive evidence about the ways in which Lucretius paints himself as a *vates* or an *augur*, but the point now has been made that Lucretius is, in a

47. See further Cicero *Att.* 13.42.3; Livy 10.37.15; Festus 146.12–17 (Lindsay).

48. The most compelling suggestion is perhaps that by speaking Lucretius renders the "sacred precincts of the mind" *liberata* ("freed"), thus allowing new ideas to take hold. Yet, as our knowledge of the ritual itself is somewhat hazy, such surmise remains uncertain. For further on the ritual, see BNP 1.22–23, Wissowa 1912: 528, and Varro *LL* 7.8–10. As *templum* is a fairly common word in Lucretius (occurring 25 times), further exploration of the implications of augural ritual for his text would no doubt prove frutiful.

central way, a speaker of power (that is, a speaker of the performative[49]), whether we call him a prophet, a priest, or neither.[50] Yet the trajectory of this introductory section, in moving from the creation of a philosophical vocabulary to the *poeta creator* to the poet as *vates* or *augur*, also makes clear a certain distance from its point of departure. Unlike Catullus' *c.* 16, which served as our model for locating the centrality of the performative in the *De rerum natura*, Lucretius' text does not "puckishly" parade before the reader the disjunct between constative and performative. Despite, or perhaps because of, Lucretius' recognition that this disjunct poses a difficulty for his project[51]—whether we take that project as merely delivering received Epicurean wisdom ("the way things are") or, with much recent scholarship, as providing therapy for his contemporaries (or, most probably, as simultaneously doing both)—Lucretius, as Dryden writes, "… seems to deal *bona fide* with his Reader, and tells him nothing but what he thinks … "[52] Lucretius' sincerity, his *bona fides*, contrasts strongly with Catullus' puckish *persona*;[53] and it, as we shall see, may play a role in his engagement with performative language.[54]

---

49. With this parenthetical comment, I do not suggest that being a "speaker of power" and a "speaker of the performative" are entirely synonymous but rather that they intersect and overlap. A "speaker of the performative" is always a "speaker of power" but a "speaker of power" is not necessarily a "speaker of the performative."

50. In this context, it is also worth mentioning the opening of Cicero's *De Fato* (1.1), which (while fragmentary) runs as follows: *quae de re futura cum aliquid dicunt deque eo* ***quod possit fieri aut non possit***, *quam vim habeant, obscura quaestio est quam περὶ δυνατῶν philosophi appellant, totaque est λογική quam rationem disserendi voco.* Perhaps Lucretius and the *De rerum natura*, which so often seems to be about "what can arise," should be placed in this larger context, a context of possibility, potentiality, and speech about the two. See further Begemann 2012: 38–47.

51. Lucretius' comments at *DRN* 1.142–45, where he (rather earnestly) describes his lucubrations on the best way to convey illumination to his reader, make clear (at least implicitly) that his performative act (of creating a new philosophical vocabulary) may go awry. That is, the stated intentions (*esse*) and the performative effects (*posse*) of his text may fail to coincide.

52. Dryden 1995: 386. See, too, Wormell 1960: 63: "I find it difficult to understand how anyone can read through Lucretius' poem without coming to the conclusion that sincerity and integrity are two of the keynotes of his character."

53. At *DRN* 1.50–53 Lucretius (more or less explicitly) proclaims his *bona fides* to the reader: *Quod superest, vacuas auris animumque sagacem/ semotum a curis adhibe veram ad rationem,/ ne mea dona tibi* ***studio*** *disposta* ***fideli,***/ *intellecta prius quam sint, contempta relinquas.* ("Furthermore, apply your unoccupied ears and keen mind,/ removed from cares, to the real account/ lest you leave my gifts, given to you with **faithful zeal,**/ scorned before they are understood.")

54. See Kennedy 2000b (on Ovid): "The trick of teaching … is one of *confidence* [emphasis added] …"

## II. *Fides, Foedus, and the Lucretian Promise*

*"fides" enim nomen ipsum mihi videtur habere, cum fit quod dicitur.*[55]

Having now, at least in limited fashion, motivated the notion that performative language plays a central role in the *De rerum natura*, we turn to examine a specific performative, the promise, as well as some terminology intimately bound up with that performative.[56] The discussion here and in ensuing chapters makes no claim to be comprehensive: Austin's own lengthy lists of the various different types of performatives give rise to the suspicion that it may in fact be impossible to enumerate all performatives.[57] Our earlier discussion of the central role that "context" plays in performative language should, moreover, make us wary of generalizing about any particular performative far beyond the local level of interpretation.[58] Nonetheless, the language of promising (which occupies—from Austin onward—a central place in the study of speech acts[59]) and, to be a bit more specific, the terms *fides* ("trust, trustworthiness") and *foedus* ("compact, agreement") hold a significant place in the lexicon of the *De rerum natura*: we have already seen one instance in which *foedus* is used to figure what are usually called the "laws of nature" (*foedera naturai*); and numerous scholars have noted that Lucretius uses "political" language (often the language of compacts, agreements, promises) to render the interaction of atoms as well as the physical world more generally.[60] Taking advantage of these results, we will now consider, through study of the vocabulary of promising, some implications of this "politicizing" of the natural world.

The chief claim of the present discussion is that a study of promises and related terminology allows us to grasp better the figuration of apparently contradictory impulses of the *De rerum natura*. In particular, an inquiry that

55. Cic. *DRP* 4.10 [7] (Non. p. 24,11). ("Fides" seems to me to have its very name, when what is said happens.)

56. To be clear: it is not the intention of this section to consider every use of *fides* or *foedus* (or every promise) within the *De rerum natura*. The discussion aims only to consider these terms in their intersection with performative language, a task less strictly defined.

57. See Austin 1975: 153–64.

58. On "context," see the second section of the first chapter of the present study.

59. See, e.g., Searle 1969: 54: "I shall take promising as my initial quarry, because as illocutionary acts go, it is fairly formal and well articulated . . ."

60. See most notably Davies 1931–32, Cabisius 1985, and Fowler 1989.

combines recognition of the performative nature of promises with awareness of their constative claims allows us to see more easily why Lucretius renders the material world using, at least in part, the language of promises. Lucretian promises, as we shall examine, raise two, if not only two, questions. The first may be called the question of materiality: Lucretian promises—in figuring our relation to sensory evidence—raise concerns about (the linguistic representation of) material boundaries, about how language names and divides the world. The second may be called the question of agency: Lucretian promises invite interrogation of what it means to act and to describe action in the world. While no discussion can promise sincerely to give the final answer to these questions in their most general form, we can put ourselves in a position to recognize better what Lucretius' text does with them.

## (i) Mutual Separation: Distinguishing the Senses and the Material World

This section focuses on a single, in many ways quite obvious, aspect of promises: they occur between two individuals, two parties, two entities of some sort. While one may make a promise to himself or to a large group, both Latin and English articulate this relationship as between a promising subject and the audience to whom he makes his pledge. The Latin idiom *fidem interponere* ("to pledge one's word," *OLD*[2] s.v. *interponere* 6b), a common way of describing a promise, expresses this idea with a local metaphor. It employs a verb, *interponere*, that elsewhere means actively "to place between or among" (*OLD*[2] s.v. 1) or passively "to lie between" (*OLD*[2] s.v. 1b). In each of these renderings, "between" translates the force of the Latin prefix *inter-*, which connotes a shared relationship.[61] This mutual give-and-take of promises is further emphasized by Ennius (*Ann.* 32 [Skutsch]), who comments in paradigmatic fashion: *accipe daque fidem foedusque feri bene firmum*. ("Give and receive *fides*, and strike a quite-firm pact.")[62] Recent Catullan scholarship, too, has noted this feature of promises.

Several Catullan poems promise, or lay claim to the rhetoric of promises. William Fitzgerald reads some of these poems against what he calls the

61. See Fest. 74.3–5 (Lindsay): **Foedus** *appellatum ... quia in foedere interponatur fides*. ("*foedus*: so named ... because in an agreement *fides* is pledged.")

62. Skutsch comments *ad loc.* (1985: 191–92) that *fides* signifies "a promise at the establishment of an association or conclusion of a treaty, thus often with *accipere* and *dare*," clearly emphasizing the mutual nature of the compact. See Sall. *Cat.* 44.3, Liv. 22.22.16.

"language of aristocratic obligation," a language that (among other things) emphasizes reciprocity.[63] A few words from R. O. A. M. Lyne may help to delimit this sphere of discourse:

> Rome was in essence an aristocratic society and possessed an elaborately developed code of social conduct.... Virtually every page of the correspondence of Cicero (to name one obvious source) attests both code and terminology. I mention here simply the essential elements. *Fides* ("fidelity," "integrity") was, or should be, the foundation of all actions and relationships; one conducted oneself in accordance with *pietas* ("sense of loyalty," "conscientiousness"). One had a profound sense of *officium* ("service," "dutifulness"); one was pleased, indeed compelled, to find and to display *gratia* ("favour"). Embracing and applying all these and other ideas was the extensive, sometimes very formal relationship connoted by *amicitia*: a complex and profound "friendship" implying at best *mutual* obligation, *mutual* affection, and *mutual* pleasure.[64]

While the broad-scale generalization Lyne employs here inevitably simplifies the evidentiary details of aristocratic interpersonal relations, his description—particularly his emphasis on the mutual (if not equal) nature of *amicitia* ("friendship")—is apt.[65] It may largely be verified by glancing at a letter of Cicero (cited by Fitzgerald):

> Breve est quod **me tibi** praecipere **meus** incredibilis **in te** amor cogit: tanta est expectatio vel animi vel ingeni tui, ut **ego te** obsecrare obtestarique non dubitem *sic* ad nos confirmatus revertare ut, quam

---

63. I have some small misgivings about classifying this terminology as *solely* the "language of *aristocratic* obligation," particularly as some of the central terms in this sphere of interpersonal relations were also used to name (among other things) international relationships. (One can, for example, see the broad semantic range of *fides* in Freyburger 1986.) Although we may suggest that international relations (and its language) are only an extension of aristocratic obligation, it seems unclear that the vectors of linguistic influence flow in *only* one direction; that is, it is unclear that the behavior of nations always was modeled on the behavior of aristocrats and never *vice versa*.

64. Lyne 1980: 24–25. Emphasis added.

65. *Amicitia* has long been an important topic of discussion in writing on Roman literature and culture, and the degree to which it actually embodied "friendship" (as opposed to political alliance) is often debated. A full bibliography would be enormous, but for general historical discussion, see, e.g., Brunt 1988b and the various remarks of Hellegouarc'h 1963 (*passim*). For consideration of the philosophical tradition on friendship, see Powell 1990 and Konstan 1997 (with his bibliography): 108–12 (on Epicurean friendship) and 122–48 (on Roman friendship, political and philosophical).

> expectationem tui concitasti, hanc sustinere ac tueri possis; et quoniam **meam tuorum erga me** meritorum memoriam nulla umquam delebit oblivio, te rogo ut memineris, quantaecumque tibi accessiones fient et fortunae et dignitatis, eas te non potuisse consequi ni meis puer olim fidelissimis atque amantissimis consiliis paruisses. Qua re hoc animo in nos esse debebis ut aetas nostra iam ingravescens in amore atque in adulescentia tua conquiescat. (Cicero *ad Fam.* 2.1)
>
> Brief is that which **my** unbelievable love **for you** compels **me** to advise **you**: so great is the expectation for your mind and spirit that **I** do not hesitate to ask and beseech **you** to return to us so strengthened that you can sustain and guard this expectation which you have stirred up; and since no oblivion will ever destroy **my** recollection of **your** services towards **me**, I ask you that you recall, whatever benefits of fortune and honor will arrive for you, that you would not have been able to achieve these if you had not formerly obeyed my most faithful and loving advice. Your duty will be, then, to be in this disposition so that our age, now growing heavy, may come to share in the respite of your love and early manhood.

Here, as Fitzgerald remarks, "[t]he prose unfolds as a set of variations on the relation between first and second persons, a configuration that recurs obsessively in an almost kaleidoscopic series."[66] The boldface type here highlights only some of the most obvious instances of the extensive coupling of "you" and "I"; the text simply overflows with them. Cicero's rhetorical (and semantic) emphasis on reciprocity is almost mind-numbing. And yet, as Fitzgerald is well aware, Cicero's text provides an insightful, if hardly neutral, backdrop for Catullan poetics.

Consider the following poem:

> *Nulla potest mulier tantum se dicere amatum*
> *vere, quantum a me Lesbia amata mea est.*
> *nulla fides ullo fuit umquam foedere tanta,*
> *quanta in amore tuo ex parte reperta mea est.* (c. 87)
>
> *In truth, no woman can say she was loved as much*
> *as my Lesbia was loved by me.*
> *No fides in any compact was ever as great*
> *as was found on my part in my love for you.*

66. Fitzgerald 1995: 129.

In analyzing this poem, Fitzgerald remarks:

> The repeated protestation of faithfulness, and the parataxis of the two audiences they imply (the general public and Lesbia), bespeak the isolation of this compact, which is not part of any network; what we hear in the symmetry and repetition of the couplets is the frustration of a voice that can only rearrange the same pieces and repeat them, a voice that has to address each audience afresh because there is no world that connects them.... Catullus' use of the language of aristocratic obligation is perverse because he gets stuck in the pose of outrage, making an absolute of values that are essentially to do with the arrangement and rearrangement of alliance; he has fallen out of circulation.[67]

Fitzgerald's reading here depends, at least in part, upon knowledge of the workings of promises. He reads Catullus against the paradigm of the mutual compact, a paradigm of reciprocity provided by Cicero and the language of aristocratic obligation. Unlike the Ciceronian ideal, Catullus' compact, his promise, in *c.* 87 is not reciprocal. Catullus has been left in the lurch; he has "fallen out of circulation"; he desperately conjures with a useless language of obligation that does not function within his isolated, amorous context. To harken back to the Ennian passage cited at the outset of this discussion, *fides* here has been given (*data*) but not received (*accepta*).

Shoshana Felman is perhaps the literary scholar who has theorized the mutual or not-quite-mutual nature of promises most extensively. With a few necessary qualifications, we may say that Catullus here succumbs to what Felman has called the "seducer's strategy" (paradigmatically, for Felman, the strategy of Molière's Don Juan):

> ... the seducer's strategy is paradoxically to create, in a linguistic space that he himself controls, a *dialogue of the deaf*.... Although he has no intention whatsoever of keeping his promises, the seducer, strictly speaking, does not lie,[68] since he is doing no more than playing on the self-referential property of the performative utterances, and is

67. Fitzgerald 1995: 131–32.

68. We might say that the seducer (forever) does not lie *yet*. What Felman points to is, in part, a temporal issue: promises (which literally "send" [*mitt-*] something "forth" [*pro-*] into the future) are always taking advantage of the fact that statements about the future cannot be purely constative, cannot describe the world in terms that may *yet* be judged true or false.

> effectively accomplishing the speech acts that he is naming. The trap of seduction thus consists in producing a *referential illusion* through an utterance that is by its very nature *self-referential*: the illusion of a real or extralinguistic act of commitment created by an utterance that refers only to itself.[69]

Felman here points to a feature of promises—in fact, a feature of performative language more generally—that we have emphasized in the first chapter: performative utterances are self-referential. In earlier discussion, we drew upon the work of Émile Benveniste, who was perhaps the first to articulate this peculiar linguistic property of performatives. Without entirely re-iterating Benveniste's claims, we may note that his analysis suggested that the performative was self-referential because it was a unique act. Thus, to choose a common example, the marriage rite, although based in and upon a standard (that is, recognized and recognizable) ritual context, is undeniably unique, since in every felicitous instance it creates a new marriage. Moreover, insofar as it refers to the act it performs, each performative utterance creates its referent, and is therefore self-referential.

Applying this analysis to the promise, a specific type of performative, allows us to see Felman's point. As a performative act, the promise may only be evaluated in terms of its felicity, not its veracity. Yet because of its mutual nature, its perceived status as a tie or bind, the promise constantly creates constative expectations: the *bona fide* promise always creates the expectation that it has authorized a corollary constative utterance. Thus, returning to Catullus *c.* 87, we see that the speaker—while not actively employing the "seducer's strategy"—is nonetheless seduced by his own promises. As Fitzgerald emphasizes, Catullus tries to call upon the mutual, reciprocal nature of the language of aristocratic obligation, but there is nothing reciprocal about the promise Catullus invokes. The *fides*, the promise, is entirely on Catullus' side (ex parte ... *mea*).

This Catullan background provides a foil for reading Lucretius' most extensive engagement with the term *fides*. As mentioned previously, Lucretius—at a certain remove from Catullus' amorous delusions—has transferred the language of promises, the language of *fides*, to the material world. Specifically, Lucretius, in a lengthy passage from his fourth book, uses *fides* to

69. Felman 2002: 17.

figure our own relation to our senses and the information they provide about the material world:

> *cetera de genere hoc mirande multa videmus,*
> *quae violare* ***fidem***[70] *quasi sensibus omnia quaerunt—*
> *nequiquam, quoniam pars horum maxima fallit*
> *propter opinatus animi quos addimus ipsi,*
> *pro visis ut sint quae non sunt sensibu' visa.*
> *nam nil aegrius est quam res secernere apertas*
> *ab dubiis, animus quas ab se protinus addit.*
> *denique nil sciri siquis putat, id quoque nescit*
> *an sciri possit, quoniam nil scire fatetur.*
> *hunc igitur contra mittam contendere causam,*
> *qui capite* ***ipse sua*** *in statuit vestigia* ***sese.***
> *et tamen hoc quoque uti concedam scire, at id ipsum*
> *quaeram, cum in rebus veri nil viderit ante,*
> *unde sciat quid sit scire et nescire vicissim,*
> *notitiam*[71] *veri quae res falsique crearit,*
> *et dubium certo quae res differre probarit.*
> *invenies primis ab sensibus esse creatam*
> *notitiem veri neque sensus posse refelli.*
> *nam* ***maiore fide*** *debet reperirier illud,*
> *sponte sua veris quod possit vincere falsa.*
> *quid* ***maiore fide*** *porro quam sensus haberi*
> *debet? an ab sensu falso ratio orta valebit*
> *dicere eos contra, quae tota ab sensibus orta est?*
> *qui nisi sunt veri, ratio quoque falsa fit omnis.*
> *an poterunt oculos aures reprehendere, an auris*
> *tactus? an hunc porro tactum sapor arguet oris,*
> *an confutabunt nares oculive revincent?*
> *non, ut opinor, ita est. nam* ***sorsum*** *cuique potestas*
> *divisast, sua vis cuiquest, ideoque necesse est*

---

70. See *DRN* 1.423 for a similar use of *fides*.

71. *Notitiam* here and *notitiem* three lines later are generally (although Glidden 1985 disputes the claim) taken as Latin equivalents of the Epicurean technical term *prolepsis*, which was discussed (in limited compass) in the first chapter. The usage here, explicitly asserting the need for a previous perception (... *cum in rebus veri nil viderit ante* ... "... since he previously has seen nothing real in things ...") of what is *verum* ("real, true"), is in line with the analysis that *prolepseis* would serve as the basis of constative language but also as the condition of possibility of performatives.

*et quod molle sit et gelidum fervensve* ***seorsum***
*et* ***sorsum*** *varios rerum sentire colores*
*et quaecumque coloribu' sint coniuncta videre.*
***sorsus*** *item sapor oris habet vim,* ***sorsus*** *odores*
*nascuntur,* ***sorsum*** *sonitus. ideoque necesse est*
*non possint alios alii convincere sensus.*
*nec porro poterunt ipsi reprehendere sese,*
***aequa fides*** *quoniam debebit semper haberi.*
*proinde quod in quoquest his visum tempore, verumst.*
*et si non poterit ratio dissolvere causam,*
*cur ea quae fuerint iuxtim quadrata, procul sint*
*visa rutunda, tamen praestat rationis egentem*
*reddere mendose causas utriusque figurae,*
*quam manibus manifesta suis emittere quoquam*
*et violare* ***fidem primam*** *et convellere tota*
*fundamenta quibus nixatur vita salusque. (DRN 4.462–506)*

*We see wondrously many other things of this sort,*
*which all seek to violate our* ***trust,***[72] *so to speak, in our*
*senses—(sc. but) in no*
*way (sc. do they violate that trust), since the greatest*
*part of these deceive*
*because of mental conjectures, which we ourselves add,*
*so that what has not been perceived by sense stands for what has;*
*for nothing is more vexing than separating obvious things from*
*doubtful ones, which the mind straightaway contributes from itself.*
*Moreover, if anyone thinks that nothing is known, he does not know*
*whether even this (sc. that nothing is known) can be known,*
*since he admits*

72. In this passage, I translate *fides* more or less consistently with "trust." The important point for the argument is that *fides* occurs *between* two entities, e.g., the mind and the senses. But it should be noted that *fides* has a broad semantic range. Eduard Fraenkel (*TLL* 6.1.1.661–62), for example, classes all uses of the term under two chief headings: (1) *fides est id cui confidi potest* ("*fides* is that in which one can trust") (2) *fides est actio sive facultas confidendi* ("*fides* is the activity or faculty of trusting"), suggesting that the second (active) sense did not arise until the time of Cicero and the flourishing of rhetorical training. This historical development has been contested, most notably by Heinze 1929 (followed by Freyburger 1986), who suggests that the active sense is present from a quite early stage. However we configure the history of the word, though, the central notion of "betweenness" remains throughout.

*that he knows nothing. I shall, then, refrain from making a case*
*against this man,*
*he, who has placed himself, with his head, into his own footprints.*
*And nevertheless, even conceding that he knows that, I shall ask*
*this very question: since he has previously seen nothing real,*[73]
*whence does he know what it is to know or not know in turn,*
*which thing has created a notion of real or illusory,*
*and which has approved the difference between sure and doubtful.*
*You will find that a notion of the real was created*
*by your first sensations, and these sensations cannot be refuted;*
*for that sort of thing, which can of its own accord conquer the illusory*
*with the real, ought to be given* ***greater trust.***
*What, then, ought to have* ***greater trust*** *than the senses?*
*Or will reasoning, having arisen from false sensation, hold*
*sway against the senses, though reason itself arose entirely from the*
*senses?*
*If sensations are not real, reason, too, becomes entirely illusory.*
*Or will ears be able to reproach eyes, or touch the ears?*
*Or will taste, then, convict this touch, or noses refute, eyes conquer?*
*It is not so, I think: for power is allocated to each* ***separately****,*
*his own power for each, and therefore it is necessary*
*that we sense* ***separately*** *both what is soft, icy, or boiling,*
*and the varied colors of things; and* ***separately****, too, we*
*see whatever is conjoined with these colors.*
*Likewise, taste has a* ***separate*** *power; odors arise*
***separately****; sound is* ***separate****. And therefore it is necessary*
*that some sensations could not refute others.*
*And, moreover, they will not themselves be able to reproach themselves,*
*since they should always be held* ***equally trustworthy****.*
*Accordingly, what is seen at each moment through sensation is real.*
*And if reasoning will not explain the reason*
*why things which close up were square, far away*

---

73. I consistently render the distinction *verum-falsum*, traditionally translated by "true-false," as "real-illusory" or "real-not real." My aim in this minor change, aside from bringing out the connotations of *verus* (cf. *OLD*² s.v. 1), is to resist thinking of the issue under discussion as a propositional, logical one. I have already suggested in the first chapter (through a reading of some remains of Epicurus) that what is at stake in (at least some discussions within) Epicurean epistemology may be whether or not things are true qua act not qua propositional logic. One could, of course, retain the translations "true" and "false" and maintain that position, but the minor shift in translation is intended to signal my resistance to reading Epicureanism in overly logical terms.

> *seemed round, nonetheless it remains for the one in need of*
> *an account to give the causes for each figure imperfectly,*
> *rather than anywhere to dismiss the obvious from his hands*
> *and to violate his* ***initial trust*** *and remove all the fundaments*
> *upon which life and health rely.*

This passage, while it represents a significant departure from Catullus' amorous contexts, nonetheless draws upon much of the same language. *Fides* occurs five times over the course of fewer than fifty lines, and one gets the sense that Lucretius, no less than Catullus, is conjuring, trying to use the familiar language of personal and political obligation to exorcise his skeptical demons.[74] What is perhaps most striking about the passage as a whole is its emphasis on division. The senses are separated from the error of the mind, just as the senses are separated from each other. And this fact is one, as we have noted, that draws support from the study of performative language.

Compacts and promises, which, as this section has explored, are mutual and are often thought to bind, here serve rather to introduce division.[75] This observation is not so surprising if we recall the self-referential nature of performative utterances. As we saw earlier, promises do not, linguistically speaking, create a formal bond. Instead, they—like all performatives—name a specific, unique act, an act that in this instance presumes division as its condition of possibility.[76] In other words, it is not possible to promise without a

---

74. The *quasi* (4.463) may indicate that Lucretius perceives his use of *fides* and the language of promising to figure the relationship to our senses as quite bold. Here, as elsewhere, Lucretius seems to be, if not the first, among the first to figure both the relations within the material world and our relations to that world with the political language of interpersonal relations. See, e.g., Davies 1931–32: 36 on *connexa*.

75. It has been suggested to me that the act here figured between the mind and the senses is not promising but witnessing. I am not sure whether this is in fact the case, but the Latin vocabulary certainly collapses any terminological distinction. Moreover, as I am interested here in making a *structural* claim (i.e., that the structure of a promise creates an implicit division between mind and senses), for my argument it is sufficient to note that witnessing and promising have the same "dividing" structure. In other words, you need someone to whom you bear witness no less than to whom you promise. (We should perhaps also recall that bearing witness is at bottom a form of promising: the witness [at least in the stereotyped American legal scenario] promises [or "solemnly swears"] "to tell the truth, the whole truth, and nothing but the truth." See Derrida 2005: 75–77.)

76. Indeed, it seems to be a general feature of performatives, embedded as they often are in ritual language, that they require witnesses, that is, they require some kind of audience in order to be valid speech acts. See Miller 2005: 205, with its references to the impossibility of Wittgensteinian "private language." (On "private language," cf. Wittgenstein 1997: 94 [sec. 269] and 96 [sec. 275].)

"context," which must include someone or something to whom the promise is made. Lucretius here uses the underlying, if in some ways obvious, implicatures of any promise to sidestep a troubling ambiguity that haunts his text: ... *nil aegrius est quam res secernere apertas/ ab dubiis, animus quas ab se protinus addit.* (467–68: "... nothing is more vexing than separating obvious things from doubtful ones, which straightaway the mind adds from itself.") Any good Epicurean knows that sense perception is reliable, but where is the line between sense perception and interpretation?[77] The repetition of *fides*, taken together with the six instances of *sorsum/seorsum* ("separately") makes it seem that this passage desperately tries to create the neat divisions (between the senses and the mind, between the individual senses) it mentions rather than merely describing them.[78]

The colorful figure of the "upside-down back-to-front" skeptic (4.469–77) presents the indistinction between mind and matter (or interpretation and perception) raised and uncomfortably dismissed in this passage.[79] The traditional manner of understanding the imagery here described is that the unnamed skeptic, who denies the possibility of knowing anything, refutes himself by putting his head where his feet should be, thus giving him no basis for his claims, no "ground upon which to stand."[80] I have no interest in revising this interpretation of the passage, but I wish to highlight its emphatic focus on "the self." Line 472 (*qui capite* ***ipse sua*** *in statuit vestigia* ***sese***: "who, **himself**, put **himself** into **his own** footsteps with his head"), with its pointed redundacy, is quite vehement, and it again reminds us that the central issue of the passage precisely concerns drawing the material boundaries of the self (both inside, between the senses, and outside, separating the mind from the senses). If we recall that the term *vestigia*, at least in the *De rerum natura*,

77. On the senses as a "liminal point" for our knowledge of the world in Epicureanism and Lucretius, see Lehoux 2013.

78. If we recall the various points made in the preceding section—that Lucretius depicts himself as *poeta creator, vates*, and *augur*—then it is hardly far-fetched to think of his figuration of our relation to our senses as an act of creation.

79. On these lines, see Burnyeat 1978, who, among other things, identifies the passage as describing a traditional figure for self-refutation.

80. While it is impossible to know the specific identity of the skeptic here discussed, there are strong parallels to the present passage at Cic. *Luc.* 23.73–74: *Chius Metrodorus initio libri qui est de natura "nego" inquit "scire nos sciamusne aliquid an nihil sciamus, ne id ipsum quidem nescire aut scire nos, nec omnino sitne aliquid an nihil sit."* ("Metrodorus the Chian at the beginning of his book *de natura*: 'I deny,' he says, 'knowing whether we know something or we know nothing, even that we know or do not know this very thing [*sc.* whether we know or do not know that we do not know]. We don't know at all whether it is something or nothing.")

often carries the sense of "traces of evidence,"[81] we may surmise that Lucretius' imagery carries the double implication that the skeptic not only stands on his head but also that he stands upon (and thus covers up) his evidence. By questioning the primacy of sensory information, the skeptic deprives himself of the "traces" he is supposed to read in the world around him.[82]

In concluding his attack on the skeptical questioning of assumed divisions and boundaries (an attack that—as we have been examining—variously exploits features of promises), Lucretius specifically draws attention to the role of language in figuring material boundaries, but he does so in a largely negative fashion.[83] While throughout he figures the relationship to the senses as one of trust, a promise, Lucretius denies the power of figuration to others, commenting as follows:

> *non modo enim ratio ruat omnis, vita quoque ipsa*
> *concidat extemplo, nisi* ***credere*** *sensibus ausis*
> *praecipitesque locos vitare et cetera quae sint*
> *in genere hoc fugienda, sequi contraria quae sint.*
> *illa tibi est igitur verborum copia cassa*
> *omnis quae contra sensus instructa paratast. (DRN 4.507–12)*
>
> *Not only would reason fall entire, life itself*
> *would collapse straightaway, too, unless you should*
> *dare* ***to trust*** *in your senses and avoid precarious places*
> *and all other things of this sort which are to be avoided;*

---

81. See, e.g., *DRN* 1.403–4, where, after presenting several considerations that argue for the presence of void, the poet remarks: *verum animo satis haec vestigia parva sagaci/ sunt, per quae possis cognoscere cetera tute.* ("But these small traces [*sc.* of evidence] are enough for the keen mind, through which you could come to know everything.") The imagery of the ensuing lines makes clear that the reading of the world from traces is conceived as parallel to dogs sniffing after and chasing prey. For further on *vestigia*, see Schiesaro 1990: 91–101.

82. Within the *De rerum natura*, the term *vestigia* carries at least one more association: they are the "footsteps" of Epicurus, which Lucretius regularly asserts himself to follow. See, e.g., *DRN* 5.55–58: *cuius ego ingressus* ***vestigia*** *dum rationes/ persequor ac doceo dictis, quo quaeque creata/ foedere sint, in eo quam sit durare necessum/ nec validas valeant aevi rescindere leges ...* ("whose [*sc.* Epicurus'] **footsteps** I have followed while I pursue reasons and teach with words, by which compact each thing has arisen and how it necessarily must endure in that compact and not repeal the powerful laws of time ...").

83. In light of our previous discussion of *fari*, it is worth noting that Lucretius tellingly describes the unnamed skeptic's speech with one of its linguistic relatives, *fateor* (*fatetur*, 4.470). Addressing the link between *fateor* and *fari*, Bettini 2008: 315 remarks: " 'confession' is fundamentally a mode of 'revelation,' and it is quite natural for these two actions to be designated by terms derived from the same root."

*and to follow things which are of a contrary nature.*
*Thus, all that facility with words is in vain,*
*which has been drawn up and prepared against the senses.*

It has been suggested that, at least from the perspective of the historical evolution of Latin, *credere* ("to trust") is the active verb corresponding to the noun *fides*.[84] Yet, whether that connection is accepted or not, Lucretius, all the way to the conclusion of the discussion, continues to press the language of political obligation in his figuration of our relation to our senses. His final remarks, however, form an interesting coda to the preceding discussion. To this point, in describing the senses, Lucretius has employed continually the rhetoric of trust and obligation. Yet he has not considered such description as an explicitly linguistic problem. His final jab at skepticism, though, takes precisely this tack. He proclaims that the *copia verborum* ("abundance of/facility with words") prepared against the senses is in vain, empty (*cassa*), a statement that sounds remarkably like a proclamation (performative, not constative) that any opposing (skeptically minded) speech act must be infelicitous. Whatever the case, having himself conjured for over fifty lines with promises and their various effects, Lucretius perhaps here indulges in a bit of transference (Freudian *Übertragung*), displacing onto his reader his own fears that his performative grafting of political obligation onto sense perception may prove empty.[85] These fears perhaps also account for Lucretius' noted sincerity, his bona fides to which we pointed earlier. Unlike Catullus, who can never seem to link constative and performative convincingly, who always conjures to little effect, Lucretius not only conjures sincerely but also to great effect, if not in his own time then at least in ours. The fact that we moderns have little trouble "trusting" in our senses no doubt owes something, if only indirectly, to the seduction of Lucretian promises.[86]

## (ii) Atoms and Agents: The Politics of Promising

Thus far, the discussion of how the language of promising figures in our relationship to the material world has focused largely on the effects of promising,

---

84. See Benveniste 1969: 120–21, Meillet 1922.

85. *Cassa* here seems to echo κενός in the well-known Epicurean *dictum* (discussed in the first chapter): κενὸς ἐκείνου φιλοσόφου λόγος, ὑφ' οὗ μηδὲν πάθος ἀνθρώπου θεραπεύεται ("Empty is the word of that philosopher by whom no human suffering is alleviated . . ."). This connection, if correct, would suggest that part of the attack on the skeptic position is that it is (at least in Lucretius' eyes) in no way therapeutic.

86. Terada 2005 represents an intriguing modern interrogation of the division between perception and interpretation, a division that Lucretius (as we have seen) tries so hard to establish.

that is, on the ways in which promises act upon the world—naming it, dividing it, creating the expectation of a mutual bond. Discussion now turns from these effects to the act of promising itself. In some ways, from the materialist perspective of atoms and void, any division between act and effect is artificial. As the analysis in this chapter and earlier has noted, action, whether the action of speech or otherwise, is always bound up in a context, which—even if it does not determine that action—nonetheless determines how that action is received and understood. Austin's felicity conditions, for example, remind us that if you appoint a horse *consul*, you will not, at least in any standard sense, have successfully appointed, but you will undeniably have done *something* (which we may call an unintended effect, a misfire, rather than a proper act, but this division already begins to bog us down in the murky quagmire of intention).[87] Granted that the division between act and effect is provisional, it nonetheless aids in structuring the present discussion.

In interrogating the act of promising rather than its effects, we make an argument that the well-recognized and sustained analogy between humans and atoms (which is undeniably something more than an analogy) raises questions for what it means to act in the *De rerum natura*. To be more specific, continuing our investigation of the language of promising, particularly the term *foedus*, we suggest that the paradigm of treaty-making (an act) provided by the *fetiales* (the fetial priests, who were responsible for, among other things, both making international treaties and formally declaring war) allows us, when read against the analogy between humans and atoms, to understand better the figuration of the Lucretian world through *foedera*. In particular, we suggest that Lucretius employs the term *foedus*, uniquely among his contemporaries, as part of a larger displacement of the category of the (intentional) act, in part replacing acts with events.

### *(a) Humans, Atoms; Atoms, Humans: Agency and the Failure of Analogy*

Before turning to a closer examination of *foedus* (and the typically Lucretian phrase *foedera naturae*), I want first to motivate a larger context for that investigation. Lucretius, it has been noted, variously "transfers" terms "properly" naming interpersonal relations to the sphere of material relations, most prominently to relations between atoms. But it is worth seeing just how extensive

87. See Austin 1975: 35 for his remarks on appointing a horse to the office of *consul*. The comment is no doubt a reference to Caligula's horse, Incitatus, which is mentioned at Suet. *Cal.* 55, Cass. Dio 69.

this "transfer" really is. Lucretius' atoms, for example, occasionally are presented at war:[88]

*nec superare queunt motus itaque exitiales*
*perpetuo neque in aeternum sepelire salutem,*
*nec porro rerum genitales auctificique*
*motus perpetuo possunt servare creata.*
*sic* ***aequo geritur certamine*** *principiorum*
*ex infinito contractum tempore* ***bellum:***
*nunc hic nunc illic* ***superant*** *vitalia rerum*
*et* ***superantur*** *item. . . . (DRN 2.569–76)*

*And so they are not able to conquer, the end-bringing*
*movements,*
*nor continually to bury life for all time,*
*and, moreover, the movements that increase and generate things,*
*they are unable forever to preserve the things they created.*
*So* ***it is waged, with equal struggle*** *of the atoms,*
***the war*** *established from time unbounded:*
*Now here, now there, what gives life to things* ***conquers,***
*and* ***is conquered*** *likewise.*

The language is unmistakable. In this description, atoms are constantly at war, and the generalizing tone of the passage (570: *perpetuo, in aeternum*; 572: *perpetuo*; 574: *ex infinito . . . tempore*) suggests that war is forever the way of atoms. At times, however, the attribution of war-like activity to atoms is more guarded. In the well-known comparison between atoms and motes in a sunbeam, for example, Lucretius writes:

*multa minuta modis multis per inane videbis*
*corpora misceri radiorum lumine in ipso*
*et* ***velut aeterno certamine proelia pugnas***
***edere turmatim certantia*** *nec dare pausam,*
*conciliis et discidiis exercita crebris;*
*conicere ut possis ex hoc, primordia rerum*
*quale sit in magno iactari semper inani. (DRN 2.116–22)*

88. On this passage, see Anderson 1960: 11–19, who at times overemphasizes his theme of discontinuity.

*You will see many tiny bodies variously mixed*
*through the void in the very light of the sun's rays*
*and,* ***as if with endless struggle, they produce,***
***in attacking throng by throng, battles and fights,***
*and do not cease, occupied with constant assemblies and discord;*
*so that from this you could guess of what sort are*
*the constant tossings about in the great void.*

The battle imagery is applied to atoms only by analogy; *velut* ("as if") seemingly qualifies any full comparison. But the language here, as we have already seen, often extends beyond analogy. The term *concilium* ("assembly"), for example, which often describes public meetings,[89] not only is drawn from political language but, together with the technical coinage *conciliatus*,[90] serves Lucretius as the *vox propria* for atomic compounds. Attempts to explain the relationship between human warfare and atomic motion as merely one of analogy thus seem to founder: *conciliare* ("assembling") is what atoms do.

A striking passage, where Lucretius vehemently denies that there is a finite number of atoms, even for the most unique of items (2.542: *unica res*), makes clear this central function of *concilium* and its cognates:

*quippe etenim sumam hoc quoque uti finita per omne*
*corpora iactari unius genitalia rei,*
*unde, ubi, qua vi et quo pacto congressa coibunt*
*materiae tanto in pelago turbaque aliena?*
*non, ut opinor, habent rationem* **conciliandi** *... (DRN 2.549–51)*

*And, in fact, if I shall grant this, too—that everywhere a finite*
*number of bodies generative of one thing are tossed about—*
*whence, where, by what power, and how will they*
*come together and combine in such a sea and foreign crowd?*
*They do not, I think, have a plan* ***of assembly*** *...*

89. See *OLD*² (s.v. 1): "A popular assembly, public meeting or gathering, especially that of the *plebs* at Rome ..."

90. See Fowler 2002: 180: "*Conciliatus* is found 4 times in L....; elsewhere only in Ps.-Rufin. *In Os.* 12.12 (PL 21).... It is more obviously a technical term than *concilia* ..." The four occurrences are *DRN* 1.575, 2.100, 2.134, 2.936.

This argument turns upon the fact that atoms need a *ratio conciliandi* ("a plan of assembly"). Atoms must be so plentiful that they may readily combine. The language here does not form a simile, an analogy, or even a metaphor, if by metaphor we mean that there is some other, literal way to name the atomic movement under consideration. At the same time, the political connotations of the words in this passage remain alive, not only in the sense that Lucretius uses terms that previously have other significations but also in that he deploys them in a fashion that in no way closes off those other significations. As Gail Cabisius comments,

> [t]hree verbs in the preceding lines, *congredior, coeo*, and *concilio*, usually describe meetings of people who have come together for a common purpose. This metaphor is sharply contrasted with the *turba aliena*. The atoms should not make up an aimless crowd in which all are unrelated and even strange to each other, but must act with purpose and agreement, if they are to create *concilia*.[91]

Cabisius brings out well how extensively the atoms in this passage are depicted as personal agents, emphasizing the ways in which they convene together for a common purpose. And, in many ways, this effect is the chief result of the prominence of political language within Lucretius' descriptions of the atomic world. Atoms, which may not in Lucretius' day have had any "proper" vocabulary for their description, continually end up "acting" like humans.[92]

There is also a flipside to this phenomenon. Alongside the humanization of atoms, we find the atomization of humans. In the early stages of the development of civilization within Lucretius' fifth book, Venus compels lovers to join together, lovers who "act" quite like atoms:

> *nec commune bonum poterant spectare, neque ullis*
> *moribus inter se scibant nec legibus uti.*
> *quod cuique obtulerat praedae fortuna, ferebat*
> ***sponte sua** sibi quisque valere et vivere doctus.*

---

91. Cabisius 1985: 111.

92. Sedley 1998: 38–39 discusses the various Latin renderings of "atom" (but not atomic motion). The variety of terms for atoms themselves—*corpuscula, atomi, individua*—suggests that such terminology was in flux in Lucretius' day. See, too, Reiley 1909: 35–66 and Reinhardt 2005.

*et Venus in silvis iungebat corpora amantum;*
***conciliabat** enim vel mutua quamque cupido*
*vel violenta viri vis atque inpensa libido*
*vel pretium, glandes atque arbita vel pira lecta. (DRN 5.958–65)*

*They were not able to consider the common good, and*
*they did not know how to use any manners nor laws.*
*Whatever prize fortune granted to each, he would bear*
***by his own power** for himself, having learned to live and fare well.*
*And Venus used to join the bodies of lovers;*
*for either common desire **attracted** each woman*
*or the violent force of a man and his excessive lust*
*or a bribe, acorns and arbute-berries or select pears.*

We have already seen the prominence of *conciliare*, here applied to humans, in atomic motion. *Sua sponte* ("by one's own agency, by oneself") is similarly prominent in atomic descriptions. In making his contention that this world is not unique, for example, Lucretius argues that it is highly unlikely that atoms external to our world do nothing;

*cum praesertim hic sit natura factus, et ipsa*
***sponte sua** forte offensando semina rerum*
*multimodis temere incassum frustraque coacta*
*tandem coluerunt ea quae coniecta repente*
*magnarum rerum fierent exordia semper,*
*terrai maris et caeli generisque animantum. (DRN 2.1058–63)*[93]

*especially since this world was made by nature, and the seeds*
*of things themselves, **under their own power**, by chance collisions,*
*variously driven, haphazardly, in vain, to no effect,*
*finally fostered these things which, connected, straightaway*
*become forever the beginnings of great things,*
*of earth sea and heaven and animate creatures.*

93. See *DRN* 2.1090–92: *quae bene cognita si teneas, natura videtur/ libera continuo dominis privata superbis/ ipsa **sua** per se **sponte** omnia dis agere expers.* ("Which things, if you should hold them, known well, nature/ seems free instantly, without haughty rulers, herself/ to conduct all things **of her own accord**, apart from gods.")

*Sua sponte* is here grouped with a stream of adverbs (*forte, moltimodis, temere, incassum, frustra*) that together emphatically present the (initially) undesigned state of atomic creation, a state hardly far-removed from the chaotic coupling of humans just described.

Further passages continue this atomization. "Lucretius' account of the plague at Athens," Gail Cabisius observes, "shows the destruction of a society and is comparable to the dissolution of an atomic *concilium*."[94] She supports her claim with the following passage, among others:

> *nec mos ille sepulturae remanebat in urbe,*
> *quo prius **hic populus** semper consuerat humari;*
> *perturbatus enim totus trepidabat et **unus***
> ***quisque** suum propere <pro tempore>*[95] *maestus humabat.*
> *(DRN 6.1278–81)*
>
> *And the standard manner of burial did not remain in the city,*
> *by which earlier **this people** always used to make burials;*
> *for the people, as a whole, was thoroughly stirred up and afraid,*
> *and **each one**, sad, buried his own, quickly as time allowed.*

The united *populus* splits apart into individuals, just like an atomic compound, a *concilium*, in the process of dissolution.

Evidence for the atomization of humans, like evidence for the humanization of atoms, is thus unmistakable.[96] The characterization of humans and atoms intermixes so extensively that Lucretius occasionally needs to cancel unwanted implications. Notable is the following passage:[97]

> *denique iam quaecumque in rebus cernis apertis*
> *si fieri non posse putas, quin materiai*
> *corpora consimili natura praedita fingas,*

---

94. Cabisius 1985: 118.

95. *propere pro tempore* Housman | *pro re* < . . . > Mss.

96. We may even say, in a loose sense, that the "atomization" of humans infects Epicureanism at a doctrinal level. One of the central problems for Epicurean ethics, at least in the eyes of Cicero's *De Finibus* ("On Ends"), is how to reconcile its goal, pleasure (an individual, "atomic" aim), with friendship (a collective, "compound" aim). Cf., e.g., Cic. *Fin.* 1.69.

97. See Kennedy 2002: 78–85 and Fowler 1989: 146–47 (discussing *DRN* 1.1021–28) on the attribution of agency to atoms.

> *hac ratione tibi pereunt primordia rerum:*
> *fiet uti risu tremulo concussa cachinnent*
> *et lacrimis salsis umectent ora genasque. (DRN 1.915–20)*
>
> *Finally, whatever you now perceive in things apparent,*
> *if you think it impossible but that*
> *you imagine (fingas) material bodies (= atoms) endowed*
> *with similar nature, on this logic the first-beginnings of things*
> *perish for you: <thus,> it will happen that, shaken with tremulous*
> *laughter, the atoms will guffaw and wet their faces*
> *and cheeks with salt tears.*

Although Lucretius himself constantly reasons from the visible to the invisible—we need only think of this comparison between motes and atoms—he nonetheless wants to limit the range of this type of reasoning. Atoms, he suggests, cannot entirely "act" like humans; they cannot have emotions. And in offering this qualification, he clearly points to the role of linguistic, if not solely linguistic, imagination: it matters how you fashion (*fingas*) atoms for yourself. Otherwise, atoms, like humans, will perish.

We are butting up continually against a central Lucretian insight: the linguistic figuration of reality matters because there may be no other ultimate access to that reality, particularly if what is most basic to reality is atoms that we cannot see.[98] As Lucretius once remarks (*DRN* 2.177) in the heat of refuting providentialist claims, "... I am ignorant of what the 'first-beginnings' of things are ..." (... *rerum ignorem primordia quae sint* ...).[99] That is, the same poet, who at one point promises precisely to tell us of "first-beginnings"

---

98. We have already explored in the preceding section Lucretius' struggle to use the language of personal obligation to separate clearly the mind from the senses, so "seeing" may not unquestionably determine an ultimate reality anyway. Yet, within an Epicurean framework as traditionally understood, anything we cannot safely infer from sense perception is suspect. Lucretius, of course, insists that we may safely infer the existence of atomic reality with arguments like the atoms-motes comparison; but it may be worth questioning where we draw the line between safe inference and insecure supposition.

99. The full line runs: *nam quamvis rerum ignorem primordia quae sint* ... ("For, although I am ignorant of what the 'first-beginnings' of things are ..."). Fowler 2002: 257 translates "however much I was ignorant" and suggests that the clause is counterfactual. His parallels for this counterfactual sense of *quamvis* are, however, strained and unconvincing. Neither *OLD*[2] nor Lewis-Short (s.v.) recognizes such a sense, and (at the time I write) the *TLL* has yet to reach the word. It is clear, though, that Lucretius makes this statement in the context of a "worst-case

(*primordia*) (*DRN* 1.54–55), does not know what they are![100] The aim is not so much to expose Lucretius' inability to deliver on his promises as to locate his awareness of the problem: how do we speak of that which does not have a "proper" language?[101]

Let us consider now a small piece of this larger puzzle. As we have begun to explore, the representation of humans and atoms intermingles, making the one resemble the other. Now let us investigate more specifically how Lucretius snatches the term *foedus* from religious and philosophical contexts, and how that snatching is inflected by the figuration of humans and atoms in Lucretius.[102] The figuration of atomic reality may not be a metaphor, if by metaphor we imply that there is another, literal way to represent atoms; yet—in another sense—the present investigation precisely interrogates a certain type of metaphor. Here the comments of L. Licinius Crassus, who describes the ornate style and (more specifically) metaphor as a feature of that style, are apt:

> tertius ille modus transferendi verbi late patet, quem necessitas genuit inopia coacta et angustiis, post autem delectatio iucunditasque celebravit. nam, ut vestis frigoris depellendi causa reperta primo, post adhiberi coepta est ad ornatum etiam corporis et dignitatem, *sic* verbi translatio instituta est inopiae causa, frequentata delectationis. nam "gemmare vitis, luxuriem esse in herbis, laetas segetes" etiam rustici dicunt. quod enim declarari vix verbo proprio potest, id translato cum est dictum, inlustrat id quod intellegi volumus eius rei, quam alieno verbo posuimus, similitudo. ergo has translationes quasi mutuationes sunt, cum quod non habeas aliunde sumas ... (Cicero *De oratore* 3.155–56)[103]
>
> That third manner [*sc.* of adorning an oration], that of using a word metaphorically, is common all over, which method necessity bore, compelled by poverty and narrow means, but later it was celebrated with

scenario" argument; that is, Lucretius makes a concession here that he would not necessarily make in other contexts. This recognition does not, however, change his syntax.

100. See Porter 2003: 202–3 on "body" (*corpus*), where he comments that "bodies as perceived are only metaphorically speaking bodies" (202).

101. More problematically, we may suggest that nothing has a "proper" language, but in the case of atoms, where a realm of Latin discourse is clearly under construction, we need not raise this more general question.

102. On *foedus* in Lucretius, see most recently Garani 2007 and Asmis 2008. On *foedera* in Roman poetry more generally, see Gladhill 2008.

103. This text here follows the edition of Kumaniecki (ed.) 1969.

> pleasure and goodwill: for, as clothes were first invented for fighting off the cold, then began to be worn for ornament and status, so metaphor was established due to poverty, then was attended with pleasure; for even country bumpkins say "vines 'sparkle with buds,'" "'riches' are in the grassy field," "the corn crops are 'happy.'" And when what can scarcely be named with a proper term is named with a transferred one, the similarity of that thing which we have named with a foreign word illuminates that which we desire to be understood. Therefore, these metaphors are like taking out loans, since you take what you don't have from elsewhere....

In the terms of this description, our present investigation will track some of the loans, the *mutuationes*, that shape Lucretius' figuration of material reality. Yet in Rome, as in the modern world, taking out a loan (*mutuum*) has consequences, and thus we shall be interested not only in the loan itself but also in what that loan does.

### *(b) Foedera: Compacts, Divine and Human*

What did *foedus* mean in Roman culture? First, let us consider some lexicographical, historical, and religious evidence to gain a broad sense of the term; then, we will return to Lucretius' deployment of *foedus*, looking at how he uses it in the light of contemporary philosophical and literary evidence. The larger trajectory of the discussion, as we have mentioned above, traces how Lucretius, through his use of *foedus*, displaces a traditional notion of the intentional act.

#### (1) *FOEDUS* AND FETIAL RITUAL

To begin, then, let us briefly examine some evidence from ancient commentators and lexicographers. The point of such inquiry is not so much to establish once and for all what *foedus* actually means as to look at the ways in which it was figured and perceived. This knowledge may help us to understand why Lucretius used *foedus*, as he did, to describe the "laws of nature."[104] What is so striking about this evidence is that it points, almost without fail, to consideration of fetial ritual.[105] Servius (*in Aen.* 1.62), for example, remarks: "*foedus*"... *dictum ... a fetialibus, id est, sacerdotibus per quos fiunt foedera* ("*foedus* is named

---

104. For passages that mention the *foedera naturae* ("laws of nature"), see *DRN* 1.584–98, 2.294–302, 5.55–58, 5.306–10, 5.916–24, 6.906–9.

105. While it is scarcely advisable to presume that all uses of *foedus* directly refer to fetial ritual, modern lexicographical works (no less than the ancient ones currently under examination) emphasize the centrality of fetial ritual to any conception of *foedus*. The *OLD*$^2$ entry for *foedus*,

from the *fetiales*, that is, from the priests through whom *foedera* are made"). In a slightly different (though similar) vein, Festus (74.3–5 [Lindsay]) **Foedus** *appellatum ab eo, quod in paciscendo foede hostia necaretur. Virgilius (Aen. 8.641): "Et caesa iungebant foedera porca."* ("**Foedus**: named from the fact that in making peace the victim is killed shamefully [*foede*]. Virgil (*Aen.* 8.641): "And, with the sow slaughtered, they made pacts.") Isidore of Seville (*Orig.* 10.100) records the somewhat more implausible (at least to modern eyes) but likely ritually motivated suggestion: *foedus nomen habet ab hirco et haedo, f littera addita.* ("*foedus* takes its name from *hircus* ['he-goat'] and *haedus* ['young goat, kid'], with the letter *f* added.")[106]

Not only was *foedus* etymologized through reference to fetial ritual, but the *fetiales* were also, as Lucretius' contemporary Varro makes clear, etymologized through their function as makers of *foedera*:

> Fetiales [*sc.* ita dicuntur], quod fidei publicae inter populos praeerant: nam per hos fiebat ut iustum conciperetur bellum, et inde desitum, ut f<o>edere fides pacis constitueretur. Ex his mittebantur, ante quam conciperetur, qui res repeterent, et per hos etiam nunc fit foedus ... (Varro *LL* 5.86)
>
> *Fetiales* [*sc.* are so-called], because they were in charge of the public faith (*fides publica*) between peoples: it was through them that a *iustum bellum* ("just war") was declared, and then, when the war was over, that trust (*fides*) in the peace was established. Before the war was declared, an embassy was sent to seek things back, and it is through them that a treaty is made even now ...

Varro here mentions, in addition to their office as makers of peace, that is, as makers of *foedera*, that the *fetiales* served to declare war, a fact also noted by Festus (81.14–18 [Lindsay]):

> **Fetiales** a faciendo[107] dicti: apud hos enim belli pacisque faciendae ius est.

---

for example, takes the fetial paradigm as a point of departure, beginning from the definition of a "formal agreement between states" (s.v. 1) and moving through a series of different agreements—private compacts, marriages, friendship—to the final definition of "a law or limit (imposed by Nature, Fate, or sim.)" (s.v. 5).

106. These etymologies are not all, in the terms of modern linguistics, true. But linguistic scholarship does seem to agree that *foedus* is etymologically connected with *fides*, and that both are derived from the Proto-Indo-European (PIE) root **bheidh-*. See Freyburger 1986: 81. Recent scholarship (though without unanimous agreement) has also tried to connect *fetiales* to this root. See Sgarbi 1992.

107. *faciendo* Mercklin | *ferendo* codd. | *feriendo* recc.

> **Feretrius** Iuppiter dictus a ferendo, quod pacem ferre putaretur; ex cuius templo sumebant sceptrum, per quod iurarent, et lapidem silicem, quo foedus ferirent.[108]
>
> **Fetiales**: are so-called from "making" (*faciendo*), for with them lies the right of making war and peace.
>
> **Feretrius**: an epithet of Jupiter from "bearing" (*ferendo*), because he is thought to bear peace; from his *templum* ("sacred space") they take the scepter, by which they swear, and the flint stone, by which they make treaties.

This comment, if we can trust the emendation of Georg Mercklin,[109] conceives the fetial priests precisely as "makers," agents. And it is no doubt worth looking more closely at what the *fetiales* do in order to understand better what it means for them to act.

Without reviewing all available evidence on the *fetiales*, we may nonetheless quickly gain some sense of what was involved in the performance of their two primary functions—making peace and declaring war.[110] To find our way in, we may consider the brief description of the office of the fetial priest offered by Georg Wissowa:

> Whether and under which conditions a *foedus* should be made, which penalty to demand or to pay, whether to declare war and how it should be conducted, these decisions all belong to the Roman people and its magistrates alone; but the formal act of making the compact, the demanding and paying of penalties, finally the accomplishment of the declaration of war in such a fashion that the gods themselves ensure the sanctity of the international treaty[111] and that any war arising

108. I include here not only the entry for the *fetiales* but also the following one on Jupiter, as the two are clearly closely linked. In addition, textual corruption likely arose from interaction between the two entries.

109. Mercklin's emendation seems relatively secure, unless we want to postulate further corruption in the manuscripts. As *facio* is the only verb used in explanation of the etymology, it is difficult to imagine another suggestion that would have point. The older manuscript reading *ferendo* is easily explained as a copyist's error (a *saut du même au même*), borrowing from the next entry on Jupiter Feretrius. The more recent reading *feriendo* (although it conjures up the phrase *foedus ferire*) is merely a corruption of *ferendo*.

110. See further (on *fetiales*) Rüpke 1990: 97–117, which focuses on the role of the *fetiales* in the declaration of war, and BNP 1.26–27.

111. See, e.g., Liv. 9.1.7: *quid ultra tibi, Romane, quid foederi, quid dis arbitris foederis debeo?* ("What further do I owe to you, Roman? What to our pact? What to the gods who stand as judges of the pact?") For further on the *foedus* created by the *fetiales*, see Wissowa 1912: 387–88.

> from the violation of that treaty is just (*gerecht*) in accordance with the laws of men and gods, these tasks are the special "science" of the fetial priests: the legal business between peoples was originally accomplished exclusively through the fetial priests of each party, but the substantive content of the treaties was determined by the state, in whose service they operated; for the fetial priest was *publicus nuntius populi Romani* (Livy 1.32.6: "the public herald of the Roman people") and was first called to that office through the express authorization of the empowered magistrate.[112]

What is most striking about this description and, as we shall see, most relevant to our present concern is that the *fetiales* are viewed, as the quotation from Livy suggests, as the public heralds of the Roman people. In other words, although the *fetiales* act, they never act on their own behalf; rather they always act in the name of the Roman people. Consider, for example, Livy's account of the oldest *foedus* known to him[113] (an agreement between the Romans and the Albans):

> pater patratus ad ius iurandum patrandum, id est, sanciendum fit foedus; multisque id verbis, quae longo effata carmine non operae est referre, peragit. legibus deinde recitatis, 'audi,' inquit, 'Iuppiter; audi, pater patrate populi Albani; audi tu, populus Albanus. ut illa palam prima postrema ex illis tabulis cerave recitata sunt sine dolo malo, utique ea hic hodie rectissime intellecta sunt, illis legibus populus Romanus prior non deficiet. si prior defexit publico consilio dolo malo, tum illo die, Iuppiter, populum Romanum *sic* ferito ut ego hunc porcum hic hodie feriam; tantoque magis ferito quanto magis potes pollesque.' id ubi dixit porcum saxo silice percussit. sua item carmina Albani suumque ius iurandum per suum dictatorem suosque sacerdotes peregerunt. (Livy 1.24.6–9)
>
> The *pater patratus* [one of the fetial priests][114] is made for accomplishing the oath, that is, for sanctifying the pact (*foedus*); and he does

112. Wissowa 1912: 550–51.

113. Liv. 1.24.4: ... *nec ullius vetustior foederis memoria est.* ("... and no older is the memory of any other compact.")

114. The *OLD*² s.v. *patratus* reports: "A senator appointed as 'father' of a deputation to a foreign power, for the purpose of making a contract which in civil law only a father could carry out." Others, however, feel the meaning of *patratus* is somewhat more obscure. Cf. Wissowa 1912: 551n8: "Die Bedeutung des Wortes ist nicht klar ..."

> this with many words, which are proclaimed in a lengthy chant that it is not worth repeating. Then, after he has recited the laws, "Hear me, Jupiter!" he says. "Hear me, *pater patratus* of the Alban people! Hear me, you, Alban people! As those words have been recited, first to last, from those tablets or wax without treachery, and as these have been—here, today—most properly understood, the Roman people will not be the first to depart from those laws. If it does depart treacherously by public counsel, then on that day, Jupiter, may you strike the Roman people, just as I today shall slaughter this pig; and may you smite the Romans as much more greatly as your power and ability allow." When he said this, he struck the pig with his flint. Likewise the Albans performed their chants and their oath through their magistrate and their priests.[115]

It is remarkable—even if these words are hardly the exact ones used in forming any historical compact—that this ritual is entirely conducted in the name of the Roman people. The only place where an *ego* ("I") appears is in the priest's comparison between his slaughter of a pig and the retribution of Jupiter. And this emphasis on the *fetiales* not acting for themselves but instead realizing the voice of the Roman people is replayed in accounts of their ritual declaration of war.

Two central pieces of evidence on this ritual are preserved in the mixture of texts that make up Servius' commentary on Virgil's *Aeneid*. In a scene near the beginning of the ninth book of the *Aeneid* (9.51–53), Turnus—yearning to face the Trojans in battle—calls out to his men: "*ecquis erit mecum, iuvenes, qui primus in hostem—?/ en*" ("Who will be with me, my boys, who first against our foe—? Come on!"). Turnus' speech breaks off, but the poet continues: ... *iaculum attorquens emittit in auras,/ principium pugnae, et campo sese arduus infert.* ("whirling about, he sends his javelin into the breeze, the beginning of battle, and carries himself erect from the field.") Fastening upon the words *principium pugnae* ("the beginning of battle"), Servius offers the following remark for the inquisitive reader:

> PRINCIPIUM PUGNAE hoc de Romana sollemnitate tractum est. cum enim **volebant** bellum indicere, pater patratus, hoc est princeps fetialium, profiscebatur ad hostium fines, et praefatus quaedam sollemnia,

115. That other Italic peoples also had fetial priests is well attested in our surviving sources, a fact which perhaps helps to explain the (perceived) central etymological connection between the *fetiales* and *foedus*. See, e.g., Liv. 1.32.5; Dion. Hal. *AR* 1.21.1, 2.72.2; *CIL* 10.797.

> clara voce dicebat se bellum indicere propter certas causas, aut quia socios laeserant, aut quia nec abrepta animalia nec obnoxios redderent. (Servius *in Aen.* 9.52)[116]
>
> PRINCIPIUM PUGNAE ("the beginning of battle"): this is drawn from Roman ritual practice. When **they desired** to declare war, the *pater patratus* (that is, the leader of the *fetiales*) used to advance to the territory of the enemy; and, after pronouncing certain ritual phrases, he would say with a loud voice that he was declaring war for certain reasons, either because the enemy had harmed allies, or because they would not return stolen animals, nor servants.

This comment is supplemented by Servius Auctus, whose text is transmitted together with that of Servius and who is generally considered to preserve information from earlier commentators:[117]

> post tertium autem et tricesimum diem quam res repetissent ab hostibus, fetiales hastam mittebant. denique cum Pyrrhi temporibus adversum transmarinum hostem bellum Romani gesturi essent nec inveniret locum, ubi hanc sollemnitatem **per fetiales**[118] indicendi belli celebrarent, dederunt operam, ut unus de Pyrrhi militibus caperetur, quem fecerunt in circo Flaminio locum emere, ut quasi in hostili loco ius belli indicendi implerent. (DS *in Aen.* 9.52)[119]
>
> Moreover, thirty-two days after they sought the return of items from their enemies, the *fetiales* used to throw the spear. Finally, when at the time of Pyrrhus the Romans were intending to wage war on an overseas enemy and they could not find a place where they could perform their ritual of declaring war **through the *fetiales***, they made an effort to capture one of the soldiers of Pyrrhus, whom they compelled

116. The numeration in Thilo-Hagen refers this *commentum* to line 52, but the lemma makes clear that the intended line (in Mynors' OCT) is 53.

117. C. E. Murgia (before his death) wrote to me *per litteras electronicas*: "... Servius Auctus is here an even better source than Servius, since the D comment probably copies Aelius Donatus closely, and Aelius Donatus would have copied his own sources closely. (Servius himself tends to reword, while Aelius Donatus did not.)"

118. Cf. Var. *LL* 5.86.

119. DS = Servius Auctus (a commentator on Virgil, who was—as Kaster 1988: 169 remarks—"perhaps of the seventh century, perhaps Irish"); for a succinct account of the complex transmission of Servius, see Marshall 1983. As with the previous note, the numeration in Thilo-Hagen refers this *commentum* to line 52, but the lemma makes clear that the intended line (in Mynors' OCT) is 53.

> to purchase a plot in the *circus Flaminius*, so that they might carry out the custom of declaring war as if (*quasi*) in a hostile place.

Both of these comments make clear that the *fetiales* were not perceived as simple agents in the declaration of war. In the first passage, the will (*volebant*) that sets the declaration of war in action is the plural will of individual Romans; in the second, it is *per fetiales*, through the fetial priests, that the war is accomplished, not by their agency alone.[120]

Even with only this brief description of fetial ritual, conclusions, and ones (as we shall see) suggestive for the text of Lucretius, emerge: first, *foedera*, at least those struck by the *fetiales*, are created (at least nominally) between groups, not between individual agents; second, there is something at least mildly atomic in the transactions of the fetial priests. Not only do these priests create and disband *foedera*, an activity that bears certain similarities to the ways in which Lucretius' atoms come together and form *concilia* ("assemblies") before splitting apart,[121] but it is also clear from the fetial ritual how strongly *foedera* (and their disbanding) mark boundaries and limits rather than agency. That is, as we observe in the almost comic act of compelling a captured soldier to purchase land in the *circus Flaminius* so that the *fetiales* may perform their standard ritual of throwing a spear into enemy territory, the declaration of war pointedly marks Roman and enemy, Roman and other, without however entirely making clear who, in a local, material sense, acts on either side.

### (2) *FOEDERA NATURAE*

Armed with this information about *foedera* and fetial ritual, let us now turn to investigate Lucretius' deployment of the term *foedus*. For all its significance, the term is relatively uncommon in the *De rerum natura*, occurring only twelve times. We have already seen (in our discussion of *DRN* 1.584–98) one instance of the most common manner (six instances) in which the term is put to use, in the fixed phrase *foedus* or *foedera naturae*.[122] There, the term largely

---

120. It is repeatedly emphasized throughout our sources on the late Republic that a *foedus* could not be made without the command of the people. See, e.g., Sall. *Jug.* 39.3: *senatus, ita uti par fuerat, decernit suo atque populi iniussu nullum potuisse foedus fieri.* ("The senate, as was fair, decreed that no treaty could be made without its own order and that of the people.") Similar are Sall. *Catil.* 29.3; Cic. *Balb.* 35, *Pis.* 50.

121. For *concilium* ("assembly") used to describe the material union of atoms, see, e.g., *DRN* 1.183, 1.484, 2.110.

122. I include in this count one instance (*DRN* 5.56–57) that does not explicitly mention *foedus naturae* but *foedus* alone. It clearly, however, belongs in this group.

carries the sense of a boundary or limit, although we shall examine its connotations more extensively in a moment. Other uses of the term (six in total) fall into two categories: "literal compacts," (that is, genuine interpersonal agreements, all of which occur during Lucretius' anthropology [*DRN* 5.1025, 5.1155, 5.1443]) and "metaphorical agreements" (that is, between *animus* ["mind"] and *anima* ["spirit"] or between *animae* ["spirits"] or, a one time occurrence, the "compacts of fate (*fatum*)" [*DRN* 2.254, 3.416, 3.781]). In his usage of *foedera naturae* as well as in his set of "metaphorical agreements," Lucretius has, as we shall see, made certain linguistic innovations.[123]

Together with Cicero, Lucretius is the first author in our surviving record to employ the phrase *foedus naturae*, that is, to remove *foedus* from its status as an interpersonal agreement and displace it to the *material* world.[124] Cicero uses the phrase in the remains of the *Pro Scauro*, which perhaps pre-date the "publication" of the *De rerum natura*.[125] But in at least one sense, Lucretius' usage of *foedus naturae* is distinctly more radical than that of his contemporary Cicero. Cicero's earliest use of the term (*Pro Scauro* 5) occurs in an explicitly Platonic passage, which variously recalls Plato's *Phaedo*, suggesting that all life is preparation for death.[126] He asks sarcastically: *num igitur ista*

123. The *TLL* article (6.1.1.1001–7) is quite striking. Its tripartite classification (which should perhaps be bipartite) revolves around whether a *foedus* is *inter homines* ("interpersonal") or *inter res* ("between things"), and this latter category is pioneered by Lucretius and Cicero. The third class merely offers *iuncturae notabiles*, that is, noteworthy idioms or turns of phrase in which *foedus* appears.

124. *Foedus* is not the only term for which Lucretius helps pioneer a material usage. See Davies 1931–32: 36: "Throughout the poem, Lucretius makes frequent use of the word 'nexus', and the verbs and adjectives derived from it. The generally accepted translation is the literal one of 'binding together'—used in a purely physical sense; the adjective 'connexa' is rendered by 'connected'—to possess an equivalent of this kind is the worst misfortune that can befall a Latin word. But to these translations there is, I consider, a very grave objection. In no other context in Ciceronian Latin is the sense of *material* binding found; it is always used in its legal meaning, denoting the relationship between debtor and creditor, slave and master. This use of the word is very old, dating from the Servian constitution. Yet the Thesaurus makes an exception of Lucretius, claiming that he, alone of the Ciceronian writers, uses it in the material sense. This seems to me arbitrary and unjustifiable; no reason is given, and it would be very hard to give one—except the old theory that Lucretius is dull, and does not use metaphor."

125. There is a great deal of supposition in all of this dating. Asconius (*In Scaur.*) reports that the *Pro Scauro* was delivered in the same year as the (lost) *Pro Vatinio* (54 BCE), and Lucretius' poem is sometimes dated to that year as well (based on Cicero *Q. F.* 2.10.3). Hutchinson 2001, however, has recently argued that Lucretius wrote in or after 49. Whatever the case, the two pieces of writing are roughly contemporaneous, and—although they are the first two extant sources to use the term—it is certainly possible that the term had been used prior to Lucretius and Cicero.

126. *Scaur.* 4: (*sc. Socrates disputat*) ... *hanc esse mortem quam nos vitam putaremus* ... *vitam autem esse eam cum idem animus vinclis corporis liberatus in eum se locum unde esset ortus*

*tua Sarda Pythagoram aut Platonem norat aut legerat? qui tamen ipsi mortem ita laudant ut fugere vitam vetent atque id contra foedus fieri dicant legemque naturae.* ("Your Sardinian woman there, she didn't read her Plato, or know her Pythagoras, did she? Although those men praise death, they nonetheless forbid suicide, saying it is against the compact [*foedus*] and law of nature.") In such a context, the notion of a "compact of nature" is hardly surprising. Plato, in his *Phaedo*, famously portrays Socrates as seeking to know how the (divine) mind of the world conducts all things for the best.[127] That is, in accordance with the paradigm of *foedus* observed here, *natura* is perceived as a divine agent, a party capable of establishing a compact, a law, for humans.[128]

Lucretius, however, as is well known, variously argues against the presence of a divine mind within nature, a fact that makes any analysis of his usage of *foedera naturae* more complex.[129] As Lucretius consistently personifies *natura*, there is little difficulty in accepting that he writes of *foedera naturae*, but greater difficulty arises in trying to assess and explain its significance. The phrase, although lurking throughout the poem, only occurs explicitly in six passages; and in each it is clear that the *foedera* express not so much an agreement as a limit.[130] We have already seen one of these passages (*DRN* 1.584–98); let us look briefly at the remaining five:

*nec stipata magis fuit umquam materiai*
*copia nec porro maioribus intervallis;*
*nam neque adaugescit quicquam neque deperit inde.*
*quapropter quo nunc in motu pricipiorum*
*corpora sunt, in eodem anteacta aetate fuere*
*et post haec semper simili ratione ferentur,*

---

*rettulisset.* ([Socrates argued] "... that what we think is life is death ... and life is when the same soul, freed from the chains of the body, has returned to the place whence it arose.")

127. See especially *Phd.* 97b7–99d2, where Socrates expresses his displeasure that Anaxagoras does not explain why things are for the best (τὸ βέλτιστον).

128. One may also think in this context of the Stoic divine *logos* (often portrayed as Zeus or Jupiter) that forms the counterpart to Lucretius' un-divine *vera ratio*.

129. See, e.g., *DRN* 1.1021–23, 5.419–21.

130. Asmis 2008 emphasizes the connection of *foedera* with limits (142): "Demarcated from one another by natural treaties, every created kind of thing, from entire world systems to every natural kind existing within a world, exercises powers of its own within fixed limits. Focussing on the human species, Lucretius demands that humans must use their unique natural power, reason, to recognise three basic limits: a limit to the power of the gods; a limit to their lifetime; and a limit of pleasure. These limits are enabling conditions rather than constraints, making it possible for humans to live a life of happiness."

*et quae consuerint gigni gignentur eadem*
*condicione et erunt et crescent vique valebunt,*
*quantum cuique datum est per* ***foedera naturai****.* *(DRN 2.294–302)*

*The abundance of matter was never more*
*packed together, nor with greater gaps;*
*for neither does anything increase nor fall away from there.*
*Wherefore, in which motion the bodies of first-beginnings*
*now are, in previous time they were in the same movement*
*and afterwards they will always be borne in similar fashion,*
*and which things have been accustomed to come to be, they will come to be*
*by the same agreement and they will exist and they will grow and flourish,*
*however much is granted to each through* ***the compacts of nature****.*

*cuius ego ingressus vestigia dum rationes*
*persequor ac doceo dictis,* ***quo*** *quaeque creata*
***foedere*** *sint, in eo quam sit durare necessum*
*nec validas valeant aevi rescindere leges . . . (DRN 5.55–58)*[131]

*I, having followed his (sc. Epicurus') footsteps, while I pursue*
*accounts and teach with words* ***by which compact***
*each thing has grown, how it is necessary to remain in*
*that compact and they cannot revoke the powerful laws of time . . .*

*denique non lapides quoque vinci cernis ab aevo,*
*non altas turris ruere et putrescere saxa,*
*non delubra deum simulacraque fessa fatisci,*
*nec sanctum numen fati protollere finis*
*posse neque adversus* ***naturae foedera*** *niti? (DRN 5.306–10)*

*Moreover, do you not discern that stones, too, are conquered by age,*
*lofty towers fall to ruin and rocks turn to rot,*
*temples of gods and their tired images split apart,*
*and the sanctified godhead cannot defer the limit*
*of fate, nor struggle against the* ***compacts of nature****?*

---

131. This passage does not show the full term *foedus naturae*, but its sense clearly indicates that it belongs with this group.

*nil tamen est signi mixtas potuisse creari*
*inter se pecudes compactaque membra animantum,*
*propterea quia quae de terris nunc quoque abundant—*
*herbarum genera ac fruges arbustaque laeta—*
*non tamen inter se possunt complexa creari,*
*sed res quaeque suo ritu procedit, et omnes*
***foedere naturae certo** discrimina servant. (DRN 5.918–24)*

*there is nevertheless no sign that beasts, mixed up between themselves,*
*and animate creatures with joined limbs were created,*
*because what now, too, pours forth from the earth—*
*classes of grass and fruits and fertile shrubs—*
*they are not, nevertheless, able to grow mutually joined,*
*but each thing proceeds in its own fashion, and all*
*preserve distinctions **by the fixed compact of nature**.*

*quod superest, agere incipiam **quo foedere** fiat*
***naturae**, lapis hic ut ferrum ducere possit. (DRN 6.906–7)*
*Moreover, I shall begin to lay out **by what compact of nature***
*it happens that this stone leads iron.*

No brief treatment could cover all aspects of these passages or give adequate space to the scholarly writing upon them, but it is important to draw attention to a few features. One particularly striking aspect, in the light of our preceding discussion, is how these passages both impose limits and explicitly limit or elide agency. We saw previously how treaties enacted by the fetial priests delimited boundaries, marking out Roman from enemy, Roman from other. In much the same way, Lucretius' *foedera naturae* mark out one species from the next, a fact that is particularly clear at 5.924: *et omnes foedere naturae certo discrimina servant* ("and all preserve distinctions by the fixed compact of nature"). Moreover, as with the descriptions of fetial ritual, the characterization of agency here is not one of simple, intentional action. In the first passage quoted (from the second book), Lucretius speaks of what has been given (*datum est*) without mentioning who gives, and in an earlier passage (*DRN* 1.587) we hear of what has been sanctified (*sancitum*) but not who sanctifies. Similarly, activity is denied in several other passages: it is impossible to rescind laws (*rescindere leges*: 5.58), to struggle (*niti*: 5.310), or to grow (*non ... possunt ... creari*: 5.922).

As at many other moments in the poem, it is clear that Lucretius in all of these passages takes, or portrays himself as taking, a divine or quasi-divine perspective on the world. The poet is able to go behind what a normal mortal may perceive and lay out how things "actually are." We approached this problem earlier in this chapter by speaking of "the constative of the performative"; that is, how Lucretius seeks to outline the limits on creation (on acting) but also to create those limits. Through the various guises that Lucretius takes on as poet—*vates, augur*, or otherwise—he often acts to create the world that he at times seems only to describe. We have emphasized repeatedly, in speaking of the *poeta creator* and related ideas, places where Lucretius points to his own, and our own, involvement in the creation of the universe. We may recall, for example, Lucretius scolding us not to figure (*fingas*: 1.917), not to imagine, atoms with emotions.

This background is relevant because it is possible, I suggest, to read Lucretius, at least in a structural sense, as a fetial priest.[132] Just as we have seen that the *fetiales* were the "heralds of the Roman people" (*nuntii populi Romani*), through whom war was created and peace made, so Lucretius is the crier through whom the *natura rerum*, specifically the atoms, are made to be at peace or at war. We should not forget in this connection the poem's opening invocation of Venus[133] as a counterpart to her lover Mars, a pair often read as the Empedoclean Love and Strife, but also, quite literally, war and peace, or union and destruction. Indeed, what Lucretius asks of Venus in his opening hymn is *suavis ex ore loquellas/ funde petens placidam Romanis ... pacem* (*DRN* 1.39–40: "pour forth sweet speech from your mouth, seeking quiet peace for the Romans").

This analysis also works well for explaining the agency, or the lack of agency, given to *natura*: if Lucretius is the fetial priest through whom *natura* (or *natura rerum*) acts, then—following the structural analogy—*natura* becomes the *populus Romanus*, which is certainly portrayed as an agent in fetial ritual but as a conglomerate rather than an obvious divinity. In this sense, the Lucretian answer to a Stoic or Platonic divine mind is the agency of a political collective, a collective whose very unity may require a quasi-divine act of figuration but which seems

132. This analysis is reinforced by the passage at *DRN* 1.968–83, which conjures up the throwing of the spear in fetial ritual. See Hadszits 1918: 147–48, West 1969: 46–48, and Bailey 1947 (somewhat skeptically): 766–67.

133. It is perhaps not insignificant in this context that Venus Genetrix was such a prominent Roman deity, associated both with C. Memmius and C. I. Caesar. For Caesar, we have an inscription from 48 BCE (*CIG* 2957), which describes him as τὸν ἀπὸ Αρεως καὶ Αφροδείτης θεὸν ἐπιφανῆ καὶ κοινὸν τοῦ ἀνθρωπίνου βίου σωτῆρα ("the illustrious god born of Ares and Aphrodite and universal savior of human life"). If a recent re-dating of *De rerum natura* is correct (Hutchinson 2001), then this inscription would be more or less contemporaneous with Lucretius.

far less inexplicable than an Epicurean divine agent. Such a suggestion would be reinforced by all the political language that we have already seen in describing the activity of atoms in the world: the choice of specifically political language would then serve as a clue to the larger figuration of the Lucretian world.[134]

In this connection, the comments of Hugh Sykes Davies deserve consideration:

> "Coetus," "congressus," "foedus," "imperium," "indicium," "reddo," "usurpo," all these are political and legal expressions that occur more or less frequently in Lucretian metaphors. Indeed, the practice is carried so far that one is tempted to think that it may have been conscious. It is just possible that Lucretius may have attempted to represent the machinery of the universe—the "machina mundi"—by symbols drawn from the legal and political machinery of the Republic. Such an idea is at least very typically Roman; they felt profoundly, throughout their history, that the Roman system of government had the same validity, the same unquestionable sanction that they found in the realm of Nature; sometimes it seems to them that their system was the direct outcome of the working of the natural world, that the Roman "Imperium" was no more to be questioned than the heavens, its operation no more than the mutations of the seasons. The idea finds its highest expression among the Stoics of the Republic, who so far impressed Polybius that he wrote his History according to the theory that the one aim of Providence in human affairs was the production and perfection of the Imperium Romanum.[135]

Davies paints a plausible context; however, the Lucretian framework is decidedly different—more atomic, more democratic—from that of Polybius or the Stoics. As the sarcastic remarks of third-century CE author Dionysius of Alexandria make clear, that atoms formed a democracy was a stereotype already in antiquity:

> θαυμαστή γε τῶν ἀτόμων ἡ δημοκρατία, δεξιουμένων τε ἀλλήλας τῶν φίλων καὶ περιπλεκομένων, εἰς μίαν τε κατασκηνοῦν συνοικίαν ἐπειγομένων … [Dionysius of Alexandria (ap. Euseb. *PE* 14.25 776D)]

134. Mention at *DRN* 5.7 of the *maiestas* ("majesty") of things (*rerum*) perhaps also contributes to the figuration of the atoms as the *populus Romanus*, as *maiestas* was a common attribute of the Roman people.

135. Davies 1931–32: 37–8.

> Quite wondrous is the democracy of the atoms: they greet one another as friends, and they embrace, hastening to encamp together in a single settlement . . .

Perhaps, though, it was a stereotype with some truth.

Considering Lucretius, at least in part, as a fetial priest and the atoms, the *rerum natura*, as the *populus Romanus* is thus an idea with some level of plausibility. It should perhaps, like the association of Lucretius' voice with that of a *vates* or an *augur*, not be pressed too heavily, but I want to pursue one further connection, or rather one point at which Lucretius' text displaces the paradigm of treaty-making for the *fetiales*. In our analysis of the fetial priests, we stressed how they were not themselves viewed as autonomous agents but rather their acts—declarations of war, peace treaties—were made in the name of the Roman people. And above in the *foedera naturae* passages we stressed the way in which pacts were "sanctified" without any mention of an agent. In the case of the fetial priests, however, gods acted as witnesses to guarantee their pacts. At Livy 1.24, it will be recalled, Jupiter was invited to smite the Romans (like the sacrificial sow) in the event that they departed from their agreement. Yet in the Lucretian world it seems that there are no gods that serve as witnesses (and offer retribution). The picture is murky, but, even as Venus and Epicurus are variously invoked as divinities,[136] and even as both in various ways serve as the "principals" who grant Lucretius' speech efficacy and authority, there is also a sense in which the parallel between atoms and humans seems to undermine a certain, traditional notion of a pact or promise.

The phenomenon we called the "atomization of humans" seems to undermine promises, not insofar as a promise is a performative but insofar as promises promise the constative. If we may be permitted to reach back to the text of *DRN* 1.584–98, there we spoke of the "constative of the performative," suggesting that the promise of the *foedera naturai* (and of Lucretius' presentation of them) is the promise of a pure description of what can and cannot arise. The promise of *foedus* is thus, in one sense, the promise of the divine; for divinity, as Lucretius makes clear in his fifth book (among other places), is characterized precisely by its constancy and immutability.[137] Yet for Epicureans, as we have previously emphasized,

---

136. See *DRN* 5.8 for the invocation of Epicurus as a god.

137. *DRN* 5.168–69: *quidve novi potuit tanto post ante quietos/ inlicere, ut cuperent vitam mutare priorem?* ("Or what much later change could harm those previously at peace, so that they might desire to alter their previous life?") This quality of being immutable or invulnerable was of course what made "living blessedly" (τὸ μακαρίως ζῆν: Epicur. *Ep. Men.* 127, that is, living like the blessed, immortal gods) the goal of Epicurean ethics.

the gods are off in the *intermundia* and are supposed to "do nothing."[138] In other words, although the gods exist precisely as a model of constancy and stability, the chief principle of Lucretius' world is one of change.

As we have already seen, humans, like atoms, come together and separate, and this separating, in part because of the comparison with atoms, is *naturalized* in the Lucretian world. In other words, as Lucretius proclaims at *DRN* 5.91–109, the visible world will be destroyed, and along with it, the compacts that govern the present species. The destruction of *foedera*, the rupture of promises, is, in this way, built into the Lucretian schema. And we may perhaps be able to go further: in part because of his focus on the explanatory power of atoms and void, Lucretius inaugurates a way of structuring experience that in a certain sense undermines the notion of the intentional act, or the long-term promise. In his first book, where Lucretius discusses the existence of things that are neither atom nor void, he suggests of *res gestae*, that is, of history, *merito possis eventa vocare/ corporis atque loci, res in quo quaeque gerantur* (*DRN* 1.481–82: "you could deservedly call [*sc.* it = history] *eventa* of the body and of the place in which each occurred"). The term *eventa* here, which is usually rendered with "accidents," describes the rather ephemeral existence that Lucretius gives to deeds and accomplishments.[139] Lucretius—even if he is not a reductionist who demands that all things have a cause at the atomic level—nonetheless gives such primacy of place to atoms and void that the category of "the act" begins to look somewhat different.

In one sense, the act, insofar as an act is an *eventum*, becomes more central; in another sense, the act, if it is conceived of as an intentional deed or a genuine pledge of constancy, is, if not utterly destroyed, displaced. From the perspective of atoms and void, things happen rather than people creating and causing them. Such a description may help us to see why a speech act, the promise, which acts (or perhaps better, occurs) in being said, achieves a certain prominence in the figuration of the Lucretian world. The epigraph that began this section, a maxim from Cicero rather than Lucretius, captures this sentiment quite well: *"fides" enim nomen ipsum mihi videtur habere, cum fit quod dicitur* ("'*fides*' seems to me to have its very name, when what is said happens.")[140] Here, if we understand "what is said" (*quod dicitur*) as a performative, a promise, then it has its name when it "happens," when it occurs; but

138. On Epicurean gods, see *DRN* 6.68–79, Cic. *ND* 1.45a.

139. Felman 2002: 109–10: "If the matter of History is made up, among other things, of speech acts it is because for Austin, as for Einstein, matter itself has ceased, above all, to be a 'thing': *matter itself is an event*."

140. Cic. *DRP* 4.10 [7] (Non. p. 24,11).

felicitous speech acts, of course, always do occur. The question of what they do, however, or of what agency accomplishes what they do, remains in the maxim, as in the *De rerum natura*, harder to answer.

## *III. Conclusion*

This chapter has aimed to do two things: first, it introduced and motivated the notion that performative speech formed a central category for inquiry in the *De rerum natura*; then, it proceeded to investigate a single performative, the promise, as well as two prominent terms surrounding promises, *fides* and *foedus*. This second inquiry found that the promise was significant for figuring, in the first instance, the relationship between mind and senses as well as between the senses themselves and, in the second instance, for figuring (and displacing) agency in the Lucretian world. The inquiry has aimed to demonstrate that a close study of performative language aids in understanding the figuration of the Lucretian world in both realms.

The next chapter turns to a study of the speech act of naming as well as, more specifically, to an examination of certain proper, personal names and their use in the *De rerum natura*. Naming, although somewhat arbitrarily chosen within the context of the present study, shows itself to be of importance not only in Roman culture and society but also in Lucretius' poem. Invocations by personal name within the *De rerum natura* are strikingly sparse, and the next chapter casts light upon this very scarcity.

# 3

# *Antonomasia, Anonymity, and Atoms*

## NAMING EFFECTS IN *DE RERUM NATURA*

*Poet, patting more nonsense foamed*
*From the sea, conceive for the courts*
*Of these academies, the diviner health*
*Disclosed in common forms. Set up*
*The rugged black, the image. Design*
*The touch. Fix quiet. Take the place*
*Of parents, lewdest of ancestors.*
*We are conceived in your conceits.*
—WALLACE STEVENS, "Prelude to Objects"

*nomen est, quod uni cuique personae datur, quo suo quaeque proprio et certo vocabulo appellatur.*
—CICERO De inventione 1.24 [34]

WHAT IS IN a name? Or what arises from acts of naming? According to one line of thinking, naming encompasses any act of verbal designation. Certain theories thus attempt to understand all language use as little more than a series of naming acts.[1] At its broadest, then, naming may be nearly synonymous with speaking, or indeed any language act. Whatever the validity of such theories, the present chapter takes a narrower focus: the subject under study is not all names or acts of naming in general but the proper

1. Such theories bear mentioning within the present discussion, as it has been argued that Epicurus himself authored one. Within extant scholarly discussions, Glidden 1983 most strongly takes this line. This sort of theory views language as a behavior and tries to account for that behavior (rather than, say, trying to define how each language user arrives at meaning).

name and more specifically, the effects created by certain central personal names in Lucretius' *De rerum natura*. Of course, separating off proper names from other sorts of words is a tricky business, a point exemplified well by the Hellenistic philosopher Diodorus Cronus. An anecdote records that Diodorus named his slave "However" (Ἀλλὰ μήν), an appellation for which context alone makes clear whether it is proper or not.[2] While not all proper names are so difficult to distinguish from common words, the anecdote shows that—at its limit—the definition of a proper name may be elusive. Yet the proper name also remains crucial to any sense of individual, personal identity.[3] If names are—to paraphrase Cicero in the second epigraph to this chapter—the proper, fixed terms by which each person is called, their study is, on some level, occasion for reflection about the notion of personhood and its linguistic representation. While the present discussion in no way claims to be a thorough treatment of personhood and identity in Lucretius, it nonetheless suggests that a careful analysis of passages in *De rerum natura* tells us something of value about the interplay between personal names and identity within the poem as a whole.[4]

The selection of naming as an object of study is, in some respects, quite arbitrary. But, as a reading of both the conclusion to the fourth book and several other passages scattered throughout *De rerum natura* shows, the results of such study can be related thematically both to central tropes of Lucretius' poem and to speech acts more generally. As we have already seen, scholars of the *De rerum natura* have noted numerous instances of the intermingling, at least on the level of figuration, of humans and atoms.[5] Atoms take on human characteristics, while humans, at various turns in the *De rerum natura*, assemble and dissipate like atomic compounds. This chapter suggests that

2. For this anecdote, see Ammonius *In Ar. De int.* 38.17–20. (This Ammonius passage and other related materials are collected as Diodorus frr. 6–7 Giannantoni [SSR, vol. 1].)

3. It is perhaps worth noting in this context that "identity" does not have a proper Latin ancestor (*identitas*) prior to the fourth century CE. See Souter 1949 s.v. *identitas*. One should hardly take this fact to mean, though, that there was no sense of personhood in classical Latin.

4. Although not for his use of proper names, Lucretius is often located at the beginning of historical narratives of personal identity theory. See, e.g., Alberti 1990.

5. Cabisius 1985 is a seminal essay for tracing this phenomenon. Davies 1931–32 partially anticipates her claims by exploring ways in which Lucretius depicts the natural world in political terms. Kennedy 2002: 78–85 and Fowler 1989: 146–47 both discuss the ways in which agency is (at least on a linguistic level) attributed to atoms. The textual evidence for this phenomenon is extensive. For atoms depicted in human terms, see *DRN* 2.116–22, 2.569–76 (atoms at war), 2.549–51 (atoms forming an assembly [*conciliare*]). For humans depicted in atomic terms, see *DRN* 5.958–65, 6.1278–81. The comments at *DRN* 1.915–20, which try to cancel the

Lucretius' use of proper names—the fixed, proper markers of personal identity—subtly continues this intermingling of human and atomic attributes. Specifically, the argument contends that Lucretius' use of certain proper names tends toward anonymity, creating a kind of quality-less faceless-ness reminiscent, at least in some respects, of atoms. Anonymity here combines at least two senses: first, central personal names in Lucretius are often absent or suppressed, thus rendering their bearers properly nameless; second, when they do appear, these names often create absence and distance rather than immediacy and presence, thus rendering their bearers less concretely known or knowable.

We shall turn shortly to the study of Lucretian personal names, a study located between tropes of Latin poetry and tenets of Epicurean philosophy. First, however, following a famous *dictum* of Lucretius himself—that nothing comes from nothing[6]—let us gain a sense of the cultural background (that is, the culture of Roman naming) against which his use developed. A brief survey of Roman naming practices, while inevitably partial and selective, will prepare our study of Lucretius' own practice.

## I. *Roman Names, Identity, and Deictic Naming*

On any account, it is hard to deny that personal, proper names were of great significance in Roman society. While a full examination of Roman naming would take us far from Lucretius and the poetic topics central to this chapter, it is nonetheless easy to note that names—and questions of how, when, and in what fashion they were used—are intimately bound up with questions of identity.[7] Without moving beyond terrain familiar to almost any student of ancient Roman culture, we can easily see issues of identity creeping into the ways in which Romans figured their relationship to names, specifically their elaborate system of forenames (*praenomina*), family names (*nomina*

---

implication that atoms manifest in miniature properties we see in larger (especially human) compounds, are remarkable, as they imply, if only in dismissal, a line of thinking that views atoms like miniature humans.

6. *DRN* 1.150: *nullam rem e nilo gigni divinitus umquam.* ("Nothing ever comes divinely from nothing.")

7. For a recent overview of Roman naming practices within a broad historical timeframe, see Salway 1994 (as well as the comments in the following notes). The literature on Roman names and naming practices is vast and diverse. Recent treatments (mostly from a linguistic or historical standpoint) include (in addition to Salway 1994) Gallivan 1992, Nicolet 1977, Salomies 1987, and Dickey 2002: 41–76. For a larger (if still incomplete) list of relevant works, see Dickey 2002: 46n9.

*gentilicia*), surnames (*cognomina*), as well as the occasional "added names" or "nicknames" (*agnomina*).[8] Quintilian, for one, remarks (7.3.27): "... those things are proper to the free man—what no one has unless he is free—fore-name, family name, surname, tribe..." (... *propria liberi, quod nemo habet nisi liber, praenomen nomen cognomen tribum* ...), thus linking the possession of a full set of names with free status.[9] Juvenal, speaking of the parasite Trebius, comments similarly (5.125–27): *duceris planta velut ictus ab Hercule Cacus/ et ponere foris, si quid temptaveris umquam/ hiscere tamquam habeas tria nomina....* ("you'll be dragged by your heel like Cacus, when he was stricken by Hercules, and you'll be placed outside, if you ever try to open your mouth to speak anything as if you had three names...."). In her recent commentary, Susanna Braund glosses these final words—*tamquam habeas tria nomina* (here translated by "as if you had three names")—as follows: "i.e. as if you were free, ironically, since Trebius is.... A free man had a *praenomen, nomen,* and *cognomen* ..., whereas a slave had only one name."[10] In other words, for Juvenal the connection between the mere possession of names and social status was so strong that even the offhand mention of *tria nomina* conjured up one's place in the social hierarchy.[11]

Names were not merely markers of status. As Quintilian notes, many traditional Roman names record (or at least at some point recorded) physical attributes or the circumstances surrounding birth:

> Scrutabitur ille praeceptor acer atque subtilis origines nominum: quae ex habitu corporis "**Rufos**" "**Longos**"que fecerunt ... et ex casu nascentium

8. The comments here allude briefly to the system of *tria nomina* ("three names"), which (according to recent scholarly opinion) was only in effect for a relatively brief period during late republican and early imperial times. See Salway 1994, which aims to place the practice of *tria nomina* within a broader timeframe (and therefore show its relatively limited importance, despite its prominent position in traditional scholarly accounts). Compare the remarks of Kajanto 1977: 64–67 on the chronological development of the *cognomen*, the last of the "three names" to take hold in official usage. For Lucretius, who lived during the late Republic, the system of *tria nomina* remains of importance.

9. Following Salway 1994, I treat a range of evidence (mainly from the late Republican and early imperial periods) that, while not strictly contemporaneous with Lucretius, nonetheless characterizes naming practices in effect during the composition of his poem.

10. Braund 1996: 297.

11. As Florence Dupont remarks (1992: 6), "[i]f you were a citizen, the city acknowledged it by giving you a title—your name." It is a certain irony of the Roman system of nomenclature that the *tria nomina*, at least in their earlier instances, belonged to aristocrats as well as to former slaves and new citizens (who often took two of their names, the *praenomen* and the *nomen gentilicium* from the aristocrat granting citizenship). See the discussion at Dickey 2002: 47–48.

(hic **Agrippa** et **Opiter** et **Cordus** et **Postumus** erunt) et ex iis quae post natos eveniunt, unde "**Vopiscus.**" (Quintilian 1.4.25)

The keen and exact teacher will study the sources of names: which ones they coined from appearance (*habitu corporis*), e.g. "**Rufus**" and "**Longus**"... which from the circumstance of those being born (here will belong **Agrippa** and **Opiter** and **Cordus** and **Postumus**) and which from events after birth, whence "**Vopiscus.**"

**Rufus** = "red, tawny; red-haired"
**Longus** = "tall"
**Agrippa** = "born in breech-position (feet-first)"
**Opiter** = "born with a deceased father but a living grandfather"
**Cordus** = "born late, out-of-season"
**Postumus** = "born after the father's death"
**Vopiscus** = "a twin surviving (in womb) after the death of the other twin"

Although Quintilian provides here no simple equation between name and identity—at times names designate physical features, at times events—in the passage there remains a certain implicit faith that name and identity are, or readily can be, linked. In fact, the larger context surrounding the passage shows that Quintilian took the recognition of such linkage as a fairly basic component of Roman linguistic education. Perceiving the *origines nominum*, "the origins of names," was viewed, at least in Quintilian's presentation, as a somewhat more advanced extension of learning one's nominal declensions, a lesson that all boys should take in with their early education (*in primis*, Quintilian says).[12]

Given Quintilian's easy (if not uncomplicated) association of name and identity, it is hardly surprising that negotiation between the two was a frequent poetic topic as well. The fourth-century poet Ausonius provides but one example (in plaintive elegiacs) that the name could, at least in some contexts, not only commemorate the circumstances of birth, but it could also tell the entire life path, even forecasting death. The *nomen*, that is, could be an *omen*:[13]

*Tu quoque maturos, puer immature, dolores*
*irrumpis, maesti luctus acerbus avi,*
***Pastor**, care nepos, spes cuius certa fuit res,*

12. Quintilian 1.4.22–25.

13. Similar in some respects is Horace *S.* 1.7, a poem that turns around the significance of two personal names, *Rex* and *Brutus*. Some of the force of that poem arises from the historical association of *Bruti* with getting rid of kings (*reges*). Persius, one of the characters of the poem, suggests in his closing remarks that this historical association should be an omen for Brutus in

*Hesperii patris tertia progenies.*
***Nomen, quod casus dederat, quia fistula primum***
***pastorale melos concinuit genito,***
***sero intellectum vitae brevis argumentum,***
***spiritus afflatis quod fugit e calamis.***
*Occidis emissae percussus pondere testae,*
*abiecit tecto quam manus artificis.*
*Non fuit artificis manus haec: manus illa cruenti*
*certa fuit fati, suppositura reum.*
*Heu, quae vota mihi, quae rumpis gaudia, Pastor!*
*Illa meum petiit tegula missa caput.*
*Dignior o nostrae gemeres qui fata senectae*
*Et querere meas maestus ad exsequias.* (*Ausonius Parentalia* 11)

*You, too, young boy, interrupt the sufferings*
*of old age, harsh pain for your sad grandfather,*
***Pastor****, dear grandson, whose promise was sure,*
*third offspring of a Hesperian father.*
***Your name, which chance had given, because a pan-pipe***
***first played a shepherd's hymn at your birth,***
***too late was understood as proof of a short life,***
***since breath flees reeds when played.***
*You perish, stricken by the weight of a hurled tile,*
*which the hand of a craftsman cast from the roof.*
*This hand was not the craftsman's: that hand was the*
*sure hand of bloody fate, underlying the culprit.*
*Alas, what wishes, what joys you destroy for me, Pastor!*
*I wish that roof-tile, thrown, had sought my head.*
*Would that you, who are more worthy, bemoaned the end*
*of my old age and protested sadly at my funeral!*

Ausonius thus further enlarges our sense of how proper names may be tied to identity. As we have seen, names may signify status through their mere presence and number. Now we also see that the details of names—that is, their ability at once to fulfill and transcend a simple deictic function, their ability

---

the poem: Brutus, that is, should kill Rex. (1.7.33–35: *Persius exclamat "per magnos, Brute, deos te/ oro, qui reges consueris tollere, cur non/ hunc Regem iugulas? operum hoc, mihi crede, tuorum est."*) On the *Rex/rex* pun, see Henderson 1998: 89–90.

to pick out a specific person and, at the same time, to describe that person—matter for personal identity.[14]

This briefly drawn sketch of Roman naming practices makes the claim, frequently re-iterated to this point, of a connection between names and identity, however defined, relatively uncontroversial, even if the details of that connection remain far from simple.[15] Against this backdrop, I would like to develop a naming effect to locate Lucretius' own engagement with proper names more precisely. This effect, which I call deictic naming, takes advantage of the ways in which names can pick out and point to a specific individual, whether real or imaginary, and seemingly conjure up a sense of the reality of that individual.[16] To gain a fuller sense of this notion, let us turn to Duncan Kennedy, who in his study of the discourse of Roman love elegy, describes one function of the proper noun. His description—without using the label "deictic naming" as such—nonetheless captures much of what is meant by that term:

> A so-called "proper" noun seems to demand a single point or object of reference in a way that a "common" noun does not. The derivation of the grammatical term from *proprius* ("one's own") seems to guarantee that the noun is an (or the) exclusive property of the person to whom it refers.
>
> The concept of the proper noun is also indicative of the way in which the body is so often represented as a closure on the processes

14. Intriguing for the modern who considers the complexities of Roman names are the remarks of Plutarch in the first chapter of his *Marius*, where he puzzles over what he once calls "the unevenness of custom" (ἡ τῆς συνηθείας ἀνωμαλία) within Roman nomenclature. Trying to identify the "chief name" (κύριον ὄνομα) among the three (or more) names in the Roman system seems to have been a source of frustration for Plutarch (no less than his *auctor* Poseidonios). This frustration should perhaps encourage caution in modern attempts to explain Roman naming, even ones that do not (like Plutarch) attribute undue importance to one name.

15. This sketch could, of course, be expanded in several directions. Much of the evidence used here to develop a connection between names and identity could be nuanced through comparison with the rich selection of material at Corbeill 1996: 57–98, which is assembled to examine the role of names, especially *cognomina*, in political humor of the late Roman Republic but also examines the significance of Roman names, or presumptions about the significance of Roman names, in a quite broad fashion. In particular, the sections "Names as Indicators of Character," "Living Your Name," and "Reading Names" show—in uncovering the presuppositions that allowed certain "nominal" humor—that there was a general, commonly supposed (if often incorrectly) correspondence between name and identity, or name and "character," as Corbeill often has it.

16. Deictic naming is, in a sense, a version of the common Roman presumption that names are, in various ways, tied to identity, but with deictic naming, names identify through their action, through their act of reference, rather than by description, as is often the case in traditional practice.

> of signification, taken to be undeniably a "thing-in-itself," an embodiment, indeed, of the notion that identity is at some level immanent. The use of a proper noun, a name, especially invites the kind of reading of a text that looks through it to a reality behind it.[17]

While all uses of proper nouns may not conform to Kennedy's description, he nonetheless effectively characterizes a particular function of names that frequently has captured the attention of students of Roman culture. That names can point to and pick out a specific individual has encouraged many, whether through prosopography or mere speculation, to seek out the "history" behind poetic (and other) texts. Indeed, with Lucretius himself, readers from Jerome to Cyril Bailey have been taken with constructing the poet's biography, a project that—at least on one level—may be seen as an attempt to uncover the entity, the "thing in itself," indicated by the name "Lucretius" (or more specifically and fully, the names "Titus Lucretius" or "Titus Lucretius Carus").[18]

Part of Kennedy's point in presenting the apparent ability of proper nouns to point to and pick out specific, embodied objects is to challenge a way of reading Roman love elegy—a way of reading hardly foreign to Lucretian studies—that depends upon an easy identification of the names (e.g., Cynthia [in Propertius], Corinna [in Ovid], Lycoris [in Gallus], et al.) of that poetry with some historical reality. Lucretius' contemporary Catullus is one author who frequently has enjoyed such reading. The Catullan corpus, often noted for the feeling of intimacy it creates,[19] is replete with proper names, often the names of Catullus' contemporaries, and above all with the name of Lesbia, his fictional or not-so-fictional beloved. Lesbia, once identified (as she so often has been) with the "historical" Clodia, is the name around which entire biographies of Catullus have been constructed, and we may go so far as to say that the use of Lesbia's name in an intimate and confessional way has inspired, at least in part, the whole industry of biographical criticism surrounding Catullus.

The plaints of Catullus *c.* 72 provide one example of this poetic intimacy, this intimate usage of proper names:

> *Dicebas quondam* ***solum te*** *nosse* ***Catullum,***
> ***Lesbia,*** *nec prae me velle tenere Iovem.*

---

17. Kennedy 1993: 83.

18. On Lucretian biography, see Canfora 1993, Holford-Strevens 2002.

19. Fitzgerald 1995: 212–15 discusses Catullus' history of conjuring up a sense of intimacy among his readers.

*Dilexi tum te non tantum ut vulgus amicam,*
*sed pater ut gnatos diligit et generos.*
*Nunc te cognovi: quare etsi impensius uror,*
*multo mi tamen es vilior et levior.*
*Qui potis est, inquis? Quod amantem iniuria talis*
*cogit amare magis, sed bene velle minus.* (Catullus c. 72)

*Once you used to say that* ***you*** *knew* ***Catullus alone,***
***Lesbia,*** *and you did not wish to hold Jupiter before me.*
*Then I loved you not only as the crowd loves its girl,*
*but as a father loves his children and sons-in-law.*
*Now I know you: that's why—even if I burn more vehemently—*
*nevertheless, for me, you are much cheaper and more trifling.*
*How is this possible, you say? Because such insult compels*
*the lover to love more, but less amicably.*

The intimate tone of this lament, an effect perhaps not entirely due to personal address, nonetheless draws much of its strength from the identification of the deictic "you," the *te* of the poem's first line, with the proper name of Lesbia. The power of such identification no doubt goaded Roland Austin (in his commentary on Cicero's *Pro Caelio*) into the following matter-of-fact assessment:

> That Lesbia's real name was Clodia is stated by Apuleius (*Apol.* 10). Most scholars have agreed that this Clodia was Metellus' wife, the second sister of P. Clodius.... One of the main points in the argument ... is the identification of M. Caelius Rufus as a rival of Catullus, with the result that Cicero's speech becomes valid evidence for the character and personality of Lesbia.[20]

Following this chain of identification, Lesbia points to Clodia (evidence of Apuleius[21]), Clodia is identified with the Clodia of Cicero's *Pro Caelio* (the same proper name), and we begin to have the makings of a steamy biography,

20. Austin (ed.) 1960: 149–50. For a recent exploration of Austin's practices, concerns, and influences in the composition of the commentary, see Henderson 2006: 9–36.

21. The Apuleius passage to which Austin refers runs as follows:

> hic illud etiam reprehendi animadvertisti, quod, cum aliis nominibus pueri vocentur, ego eos Charinum et Critian apellitarim. eadem igitur opera accusent C. Catul<l>um, **quod Lesbiam pro Clodia nominarit**, et Ticidam similiter, quod quae Metella

all of which is buttressed ("one of the main points in the argument") by the appearance of a "Caelius" in two poems of Catullus, *cc.* 58 and 100.[22]

While it may seem somewhat easier for present-day scholars to resist the sins of overly biographical criticism (and while it may seem that—in dredging up such concerns—the present discussion persists in beating a scholarly dead horse), Kennedy, after providing the above characterization of proper names, also cautions against a position (one he attributes to Paul Veyne[23]) that too easily presumes a boundary between naming and history, however defined.[24] That is, although modern scholars may not wish to follow Austin in his all-too-easy movement from Lesbia to (some) Clodia to the Clodia of the *Pro Caelio*, in simply and finally shutting off any interpretive link between Catullus and his historical circumstance (as mediated through Cicero) they would not only eliminate the possibility of any truly historical discourse, but they would also reify history (as that which is not behind Catullus' poetic discourse) just as much as Austin's equation of Lesbia with Clodia.

To put this point more concretely, we may observe that Catullus' extensive use of proper names is not far from the usage of supposedly more historical genres:

> Sed quoniam emersisse iam e vadis et scopulos praetervecta videtur esse oratio mea, perfacilis mihi reliquus cursus ostenditur. Duo sunt

> erat Perillam scripserit, et Propertium, qui Cynthiam dicat, Hostiam dissimulet, et Tibullum, quod ei sit Plania in animo, Delia in versu. (Apuleius *Apol.* 10)
>
> Here (*sc.* in my poems) you have noticed that this, too, was criticized, namely, that—although the boys are called by other names—I repeatedly named them Charinus and Critias. Let them accuse Catullus, then, for these same deeds, **since he used the name Lesbia for Clodia**; and Ticida likewise, since he wrote Perilla for her, who was really Metella; and Propertius, since he says Cynthia, concealing Hostia; and Tibullus, since he has Plania in his mind, but Delia in his verse.

It should be noted, of course, that this passage, although it has been used traditionally to support "historical" readings of the sort offered by Austin, appears in Apuleius as part of a defense that emphasizes the speaker's (that is, Apuleius' or "Apuleius"'s) own modesty in withholding the "real" names of his addressees in his erotic poetry. Pointing to a "truer" historical reality not represented in his poetry is thus advantageous for a speaker—one on trial for practicing magic but also for living immodestly—who is trying to minimize the perception of his own behavior as outlandish.

22. Fordyce (ed.) 1961: xiv–xviii (a much maligned, expurgated text of Catullus—see Henderson 2006: 70–110) gives a fuller assessment of the evidence available for the identification of Lesbia with Clodia.

23. The work Kennedy discusses is Veyne 1988. Specifically, Kennedy charges that Veyne "collapse[s] the historical into the aesthetic, and abandon[s] contingency for a metaphysical master term." (Kennedy 1993: 99) That is, Kennedy criticizes the fact that Veyne ultimately reads history as something "invented" rather than reading art as something that happens (within the course of history).

24. See Kennedy 1993: 96–100.

> enim crimina una in muliere summorum facinorum, auri quod sumptum a **Clodia** dicitur, et veneni quod **euisdem Clodiae** necandae causa parasse **Caelium** criminantur. Aurum sumpsit, ut dicitis, quod **L. Luccei** servis daret, per quos **Alexandrinus Dio** qui tum apud Lucceium habitabat necaretur. Magnum crimen vel in legatis insidiandis vel in servis ad hospitem domini necandum sollicitandis, plenum sceleris consilium, plenum audaciae! (Cicero *Pro Caelio* 21 [51])
>
> Now, since it seems that my speech has already emerged from the shallows and passed by the rocks, the rest of my course appears ahead, quite easy; for on the head of a single woman there are two charges of the greatest misdeeds: for gold (which it is reported was taken by **Clodia**) and for poison (which they accuse **Caelius** of having prepared for the purpose of killing this same **Clodia**). She took the gold, as you say, to give it to the slaves of **Lucius Lucceius**, through whom **Alexandrinus Dio** (who was then living at the home of Lucceius) was to be killed. A serious charge, whether for setting legates to lie in ambush or for inciting slaves to slay the guest of their master, a plan full of treachery, full of recklessness!

Without claiming that the usage of proper names here is entirely the same as in the Catullan example (*c.* 72) cited previously, we may nonetheless note that both Catullus and Cicero—in part through their detailed usage of proper names, names that seem to point to reality, whether or not they actually do—allow us to imagine a historical narrative. In this way, the study of deictic naming, a study of the perceived effects of the text rather than its truth, privileges neither the historical nor the aesthetic, instead leaving both in play. That is, the pointing effect under study is neither one that is "really there" and truly historical nor merely a meaningless, playful aesthetic effect but rather a perceived function of language, however and wherever we see that language functioning.

## *II. Epicurean Naming and Lucretian Anonymity*

Having identified this naming effect, I would now like to use it, along with other relevant evidence, to help us read Lucretius, particularly (although not exclusively) the conclusion to his fourth book and the so-called diatribe against love found there. What I hope to show, as indicated at the outset, is that the use of proper names in Lucretius, at least in the passages I shall read, tends towards anonymity. Lucretius' text, that is, often resists the use of personal names, whether in address or referentially, and when names do appear, they often fail to create any (deictic) sense of historical reference, a fact that is striking both, as we have seen, in light of Roman naming practices (that suggest a connection between name and identity) and, as we shall see, in light of Epicurean thinking about names. Furthermore, as we shall discuss in conclusion, the tendency of certain portions of Lucretius' text to present anonymous or vaguely named figures has point: it resonates not only with features of atomism (and, more specifically, the representation of atoms in Lucretius) but also with certain political aspects of the Roman literary tradition as exemplified by Cato's *Origines*.

While the conclusion of Lucretius' fourth book will serve as the centerpiece of the present discussion, to gain a fuller sense of Lucretius' practice, it will be useful to surround that centerpiece with other evidence concerning his use of proper names. We shall thus include both theoretical discussion of names within Epicureanism and discussion of Memmius, Epicurus, and the role that their names play within Lucretius' poem. It is important to note that this chapter does not present a comprehensive treatment of every proper name in the *De rerum natura*. The focus instead is on the effect of select passages (and therefore on the effect of select instances of naming) as well as upon locating these passages within an Epicurean framework. The argument does contend, however, that the passages of *De rerum natura* under consideration create a consistent effect, an effect of distance and anonymity. Furthermore, the consistency of this effect suggests that a fuller reading of other proper names throughout the poem would, if not re-iterate the details of the readings presented here, nonetheless unearth many similar concerns.[25]

## (i) Epicureanism and Deictic Naming

To begin, then, let us consider a comment attributed to Epicurus, a comment that has a surprising amount to say about Epicurean thinking on proper names:

> Ἐπίκου|[ϱ]ος τὰ ὀνόματά φη|σ[ι]ν σαφέστερα εἶναι | τῶν ὅρων, καὶ μέν|τοι καὶ γελοῖον εἶ|ναι, εἴ τις ἀντὶ τοῦ εἰ|πεῖν "χαῖρε Σώκρα|τες" λέγοι

25. As the analysis focuses on naming *effects* rather than names per se, the failure to consider every proper name in the text should in no way vitiate the interpretation given here, although consideration of further passages no doubt would enrich and complicate the present discussion. To provide only a rapid gesture toward a larger reading in line with some of the present

"χαῖρε ζῶι|ον λο[γ]ικὸν θνητόν." (Anon. *In Plat. Theaet.* 22.39–47 = Long-Sedley 19F)

> Epicurus says that names are clearer than definitions, and moreover, that it is ridiculous if anyone should say—instead of "Hail Socrates"—"Hail rational, mortal animal."

These words, preserved by the anonymous commentator on Plato's *Theaetetus*, contrast "names" with definitions, commenting that "names" are clearer than definitions. The fragmentary state of the papyrus makes it difficult to judge precisely why and in what context Epicurus made this remark, if indeed he made it at all;[26] but the ridiculous anecdote of greeting Socrates with "Hail rational, mortal animal" suggests that one of the advantages the name holds over the definition is a referential one. That is, names can more easily pick out and point to referents, while definitions, at least on one interpretation, merely or primarily relate words to other words, making clear that Socrates is neither irrational nor immortal but not (the perhaps more important point) that he is Socrates.[27] In other words, names here perform a function we identified previously as deictic naming, whereas definitions—even if they do not fail to refer—seemingly struggle to refer uniquely.[28]

On this interpretation of the fragment, proper names are barely, if at all, distinguishable from other names. Names, whether proper or common,

---

concerns, there are a few lists—the mention of Homer and Ennius (*DRN* 1.102–26), the catalogue of pre-Socratic philosophers (*DRN* 1.635–920), the list of "heroic deaths" (*DRN* 3.1024–52)—of proper names that figure importantly in the poem. Each of these lists, however, focuses upon a cast of characters that is long dead (in fact, that is the point of *DRN* 3.1024–52) and the names appear more for their renown than to conjure up any physical or historical reality. It is thus, at least in a limited sense, striking that a materialist poet, whose focus often falls upon describing the fundaments of the phenomenal world, nonetheless resists taking advantage of deictic naming, more than occasionally frustrating any attempt to give a sense of historical reference to his text.

26. Whether or not Epicurus ever uttered the exact words preserved in this fragment is of little concern for the present discussion. As further evidence in this chapter suggests, the general tenor of the comment made here fits well with Lucretius' own attitude toward names and naming, for which reason it matters little whether this fragment is of Epicurus himself or merely Epicurean.

27. In this instance, "rational, mortal animal" could perhaps refer—going beyond relating mere words—but it might refer to anyone within earshot, anyone in possession of *logos*, rather than to Socrates alone.

28. To be clear, unlike our earlier Catullan example, this single reference to Socrates does not necessarily conjure up a sense of his entire history; but it does—through effective, if limited, conjuring—succeed in providing a legitimate "historical" context, the greeting. The use of language in the fragment invites us to presuppose a Socrates with a minimum of human attributes, although it is entirely possible to say "Hail Socrates" in any number of contexts.

are judged not for their truth or propriety (i.e., their "proper-ness") but for their ability to refer successfully. In a comment on the above passage, A. A. Long and David Sedley point to this difficulty, focusing on the translation of ὀνόματα (a term that may be rendered variously as "words," "names," or "proper names"):

> ὀνόματα One would expect this to be "words," rather than "(proper) names." If so, the example, "Socrates," is unfortunate.[29]

Long and Sedley here suggest that ὀνόματα may be interpreted readily as either words or proper names, themselves preferring to understand ὀνόματα as words. Yet, if the line of interpretation indicated is correct, if ὀνόματα are clearer than definitions because they engage in deictic naming, then words and proper names are perhaps indistinguishable not because Epicurus has given us a bad example but because both words and proper names, on this view, aim at the same function—attaining pure reference.

The sense provided by this fragment—that deictic naming holds a central place in Epicurean linguistic thinking—is bolstered by a quick glance at the origins of language (as described in Lucretius' fifth book), where language seems designed to take the place of deictic pointing. There the *nomina rerum*, the "names for things," are expressly linked to the pointing of the child, and to the presence of the objects to which the child points:

> *At varios linguae sonitus natura subegit*
> *mittere, et utilitas expressit nomina rerum,*
> *non alia longe ratione atque ipsa videtur*
> *protrahere ad gestum pueros infantia linguae,*
> *cum facit ut* ***digito*** *quae sint* ***praesentia monstrent.***
> *(DRN 5.1028–32)*
>
> *And nature subdued the varied sounds of the tongue*
> *and sent them out; utility coined names for things.*
> *For a reason not far different, the very speechlessness of the tongue*
> *seems to draw children forward to gesture,*
> *when it brings about that* ***they show*** *what is* ***present*** *with their*
> ***finger.***

29. LS: 2.102 (19F, comment *ad loc.*).

What is described here is hardly full-blown language, but the juxtaposition of the pointing child and the coining of "names" (*nomina*) is striking. Though this evidence could scarcely be used to prove that Lucretius held a view of language as purely deictic, it nonetheless invites the reader to assume that deixis is one of the primary functions assumed by newly developed language.

Together with the earlier statement of the *Theaetetus* commentator, Lucretius' "primal scene" of language constructs a version, perhaps idealized, of language as functioning deictically in Epicurean thinking. What I want to note now is how often scenarios that resemble (at least in a generic fashion) this primal scene of deictic naming lurk in Lucretius' poem. If deictic naming depends upon a recreation of presence, a recreation of an originary moment of physical pointing in the full presence of a referent, it is striking how often this problem of presence recurs in various guises throughout *De rerum natura*. To choose but one example, *simulacra*, the thin films that constantly bombard us and provide us with visual knowledge of the world, are famously insubstantial, present but absent. One of Lucretius' most arresting images is that of the frustrated lover, left unsatisfied by *simulacra*:

> *Ut bibere in somnis sitiens quom quaerit et umor*
> *non datur, ardorem qui membris stringuere possit,*
> *sed laticum simulacra petit frustraque laborat*
> *in medioque sitit torrenti flumine potans,*
> *sic in amore Venus simulacris ludit amantis,*
> *nec satiare queunt spectando corpora coram*
> *nec manibus quicquam teneris abradere membris*
> *possunt errantes incerti corpore toto.* (*DRN* 4.1097–1104)

> *Like a thirsty man when he seeks to drink in his dreams and water*
> *is not given, water that could extinguish the burning in his limbs,*
> *but he seeks simulacra of water and toils in vain,*
> *thirsting as he drinks in the middle of a raging river,*
> *so in love Venus teases lovers with simulacra,*
> *and they are unable to satisfy their bodies with face-to-face regard,*
> *and they are unable to scrape away anything from tender limbs*
> *with their hands, wandering uncertain over the entire body.*

This passage, as Philip Hardie notes, "teaches a lesson about the vanity of desire," and in so doing, it reminds us of the importance of the presence-absence dichotomy we already have identified in our discussion of deictic naming: *simulacra*, though accessed through the eyes or fingertips, are simply not present enough ever to satisfy desire.[30] This scene, of course, represents a noticeable departure from the pointing child, but *simulacra* seem, in many ways, to operate like language (at least on one, rather deictic understanding), standing for and conjuring up full presence at the same time as they frustrate attempts to enjoy that presence.[31]

*Simulacra*, moreover, stand in a strange relation to language, as it seems that general concepts (*prolepseis*), to the extent that Epicureanism tolerated them, were imagistic and were therefore built upon repeated perception of *simulacra*, built, that is, upon repeated viewing of a ghostly (absent) presence.[32] However we understand this relationship, though, Lucretius—in his explorations of sense perception within the fourth book—manifests a fascination with the ways in which sense perception functions and breaks down. In the case of language and voice specifically, Lucretius offers the following explanation of the ways in which voice can splinter and variously communicate, disintegrate, and deceive:

> *Praeterea verbum saepe unum perciet auris*
> *omnibus in populo, missum praeconis ab ore.*
> *in multas igitur voces vox una repente*
> *diffugit, in privas quoniam se dividit auris,*
> ***obsignans** formam verbis clarumque sonorem.*
> *at quae pars vocum non auris incidit ipsas,*
> *praeterlata perit frustra diffusa per auras;*
> *pars solidis adlisa locis, reiecta sonorem*
> *reddit et interdum frustratur imagine verbi.* (*DRN* 4.563–71)
>
> *Moreover, often a single word, sent forth from the mouth of a herald,*
> *will excite the ears of everyone in a crowd.*

30. Hardie 2002: 151.

31. Indeed, although *simulacra* lie at the heart of all visual perception, Lucretius seems obsessed with *simulacra* that frustrate and mislead. See *DRN* 4.26–44 (ghosts), 4.129–42 (*simulacra* with no originals), 4.436–42 (images distorted in water), 4.962–1036 (dreams). Cf. Gordon 2002: 91–94.

32. See Fowler 1986: 299 on the imagistic nature of *prolepseis*, the much-debated Epicurean "concept." Asmis 1999 comments on the imagistic nature of Epicurean reasoning more generally (281): "Reasoning is an activity produced in the mind by means of atomic movements. To reason is to arrange images; and this mental power is the result of images continually impinging on the mind and establishing certain patterns within it."

> *A single voice thus splinters straightaway*
> *into many voices, since it divides itself for individual ears,*
> ***stamping** its form and clear sound onto words.*
> *But that part of the voices which does not fall upon the ears themselves,*
> *perishes, carried past and scattered in vain through the air;*
> *part, having crashed against solid ground, is thrown back*
> *and returns its sound and sometimes it deceives with the image of a word.*

The mechanical operation of language described here, one of voice stamping and disseminating its physical form for mass consumption, reads like a dispatch of official letters rather than a typical description of voice falling upon a crowd.[33] Lucretius' particular use of *obsignans*, the present participle that here characterizes the activity of *vox* ("voice"), receives its own entry in the *Oxford Latin Dictionary* as "to stamp, impress, imprint," though all other classical instances of the term are there united under the heading "to affix a seal to." If we surmise that Lucretius' use of this term is in fact not so different from other instances, we may suggest that the passage raises not only the issue of the mechanical reproduction of voice but also of the authenticity of that voice. A function of seals, though hardly the only one, is to authenticate and mark letters as genuine, to act, that is, as deictic signifiers that point (or seek to point) unambiguously to their particular, authentic origin, recreating and verifying their originary moment of stamping. Seals thus, in a particular way, return us to the same problematic (the problematic of the absent originary moment) as deictic naming more generally.

These various episodes in Lucretius' poem, taken together with the other, more theoretical moments (both in Lucretius' text and in the Epicurean tradition more generally), help create a sense of the importance of deictic naming in the Epicurean imagination. While one need not understand all Epicurean thinking on language as consistently voicing the centrality of deictic naming, the passages considered above show that—both implicitly and explicitly—it contributed, and contributed importantly, to Epicurean linguistic discussion.

## (ii) Memmius, Epicurus, and Anonymous Lovers

Given the prominence of the pair presence-absence for our theoretical discussion of Epicurean naming, we perhaps may suspect that now, in turning

33. See Holmes 2005: 541–43 for an insightful discussion of the voice (*vox*) and seal imagery in this passage.

to a closer examination of the actual use of proper names in *De rerum natura*, this same pair will retain an important position. This section considers in some detail Lucretius' mentions and invocations of his addressee, Memmius, and of his master, Epicurus, as well as the use of proper names within the famous diatribe against love that concludes Lucretius' fourth book. In each of these instances, the pair presence-absence recurs, specifically in trying to understand the presence or, more often, the absence of *proper* names.[34] This investigation is thus not, like the preceding section, a direct investigation of deictic naming, nor does the pair presence-absence obviously carry the same sense of invoking an originary scene of deictic pointing. Nonetheless, bearing in mind the role that deictic naming carries in evidence examined to this point, we shall have occasion to read Lucretius' text against an ideal of deictic naming, and Lucretius' text, moreover, often will invite such reading.

### *(a) Memmius*

Turning, then, to Lucretius' addressee, Gaius Memmius, we must remark his empirical absence from the poem. His name occurs only ten times (in over seven thousand hexameter lines of poetry), not appearing at all in half the books of the work. Even at the outset of the poem, where one might expect a greater mention of a patron (or a would-be patron, as Memmius is sometimes thought to be), Memmius is relatively absent.[35] He does not appear for the first twenty-five lines of the poem, and when he does appear, his appearance is rather indirect:

> *Quae quoniam rerum naturam sola gubernas,*
> *nec sine te quicquam dias in luminis oras*
> *exoritur neque fit laetum neque amabile quicquam,*
> *te sociam studeo scribendis versibus esse*
> *quos ego de rerum natura pangere conor*
> ***Memmiadae nostro**, quem tu, dea, tempore in omni*
> *omnibus ornatum voluisti excellere rebus. (DRN 1.22–27)*

---

34. I emphasize the word "proper" here, as the absence at stake is sometimes a question of the utter absence of a name, at other times of the absence of a proper name.

35. Compare, for example, the practice of Horace, who although generically and temporally distant from Lucretius nonetheless gives a sense of how prominent an addressee may be. The *Satires* (*S.* 1.1.1), *Odes* (*C.* 1.1.1), *Epodes* (*I.* 1.4), and *Epistles* (*E.* 1.1.3) all begin with invocations of Maecenas. (Maecenas' name, as these line numbers indicate, occurs within the first four lines of each of these opening poems.)

*You [sc. Venus]—since you alone direct the nature of things—and without you*
*neither does anything arise upon the divine shores of light,*
*nor does anything fertile or capable of love come to be,*
*you I seek as my ally in writing verse,*
*which I try to expound on the nature of things*
***for our member of the clan of Memmius**, whom you, goddess,*
*have wished at all times to excel and be adorned with all things.*

Here Memmius enters, named not specifically but generically by *gens* (clan, family).[36] The tone of this reference is perhaps not what one would expect for introducing the primary addressee of the poem: this passage, together with the rest of the first fifty lines of the poem, forms an elaborate hymn to Venus, a hymn that makes Memmius seem a secondary concern at best. It is true that Venus—in the form of Venus Genetrix (here invoked)—was affiliated with the *gens Memmia* and even appeared on its coinage, but this fact at best only tangentially conjures up the addressee of the poem, who, resting in an oblique case, seems almost a tacked-on afterthought.[37]

Whatever we make of this particular mention of Memmius, the rest of the poem provides little support for the notion that he or his name is of central concern. While we may, by inference, read Memmius into countless vague second-person imperatives or an occasional mention of *tibi*, it is hard to make much of occasional remarks such as *quae **tibi** posterius, **Memmi**, faciemus aperta.* (*DRN* 2.182: "We shall make these things clear **to you** later, **Memmius**.") Neither particularly deictic nor entirely frustrating of deixis,[38] this act of naming does little to challenge the thesis, asserted by G. B. Townend in his article "The Fading of Memmius," that Memmius effectively

36. *Memmiades noster*, the term here used to refer to Memmius, stands out as "generic," as it is the only instance of this lexical form in the poem. The remaining instances of Memmius' name occur in the vocative form of his *nomen gentilicium, Memmi* (*DRN* 1.411, 1.1052, 2.143, 2.182, 5.8, 5.93, 5.164, 5.867, 5.1282) or as a genitive, *Memmi* (*DRN* 1.42). According to Dickey (2002: 60), the *cognomen* is most common in poetic, single-name address, although Lucretius' contemporary, Catullus, also prefers the *nomen gentilicium*. See Fordyce (ed.) 1961: 342.

37. On the connection between the *gens Memmia* and Venus Genetrix, see Wissowa 1912: 292.

38. On the face of it, a reference like this one could be deictic, yet Lucretius does little to develop such a connection. Memmius is invoked multiple times (particularly in the first and fifth books), but the invocations are always bare invocations, which do nothing to provide a sense of the reality of the addressee. One gets the sense that invocations of Memmius do point, but they point at nothing.

disappears and morphs into a general reader.[39] But—whether or not we follow Townend[40]—Memmius, even when he is nominally present (that is, conjured up by direct address) is a rather thin, absent ghost of an addressee.[41]

Yet, if there is ample empirical evidence for Memmius' absence from Lucretius' text, the interpretation of his absence remains less than empirically obvious. One effect, though, of Memmius' nominal (and general) absence from the poem is that *De rerum natura* draws relatively little attention to its own (textual—that is, written and read—but also potentially oral and aural) performance.[42] While much has been written about the ways in which the poet figures his relationship to the reader—Philip Mitsis, for example, focuses on Lucretius' portrayal (at *DRN* 1.926–50; 4.1–25) of himself as analogous to a doctor duping children into drinking their wormwood and thus develops an argument about Lucretius' not entirely good-natured "didactic coercion"[43]—such interpretations tend to focus on Lucretius' explicit pronouncements about his poem to the neglect of the actual experience of reading *De rerum natura*. A full analysis of that experience would demand a more extensive treatment than the present discussion can provide;[44] but Lucretius' general reluctance to point (deictically, that is) to his addressee forms part of a larger project of rendering the space in which his text operates nebulous and indistinct, a project of anonymity that we shall continue to examine through the window of further proper names.

---

39. To be more precise, Townend (1978: 283) presents himself as offering a refinement of a thesis first offered by Ivo Bruns (in Bruns 1884). Bruns "argued that the change to the general reader was already taking place well before the end" of the first book, whereas Townend views the general reader as only fully present (and Memmius, therefore, effectively absent) in books three, four, and six. (Townend also, for similar reasons, believes—along with Bailey and others—that the first, second, and fifth books were written before the third, fourth, and sixth.) See Johnson 2000: 6–11.

40. Resistance to Townend's thesis would center, at least in part, around critique of his notion of the general reader: can we—in an age of feminist, post-colonial, and other genres of reader—still support the idea that there is a "general" reader?

41. Memmius' status as a ghostly addressee no doubt helps account for the wild diversity in interpretations of his role. He is sometimes considered a (desired) patron, at other times a negative example, at still other times (as we have just noted) a general or exemplary reader.

42. One could imagine a more extended reading of Memmius (following the lead of Culler 1981: 135–54 [esp. 152] on apostrophe) as little more than a reminder of the "now" of the poem.

43. Mitsis 1993. See, too, Keen 1985, which presents a less insidious Lucretius. Fowler 2000, in the context of discussing his notion of the "didactic plot," considers the different metaphors (especially path metaphors) that govern Lucretius' depiction of the student-teacher relationship.

44. Fowler 1995 discusses the segmentation of Lucretius' text and the manner in which the poet writes and groups arguments, making perhaps the greatest strides in studying the phenomenology of reading Lucretius.

### *(b) Epicurus*

Turning our attention from Lucretius' addressee to his philosophical master, we observe that Memmius is hardly alone in his nominal absence from Lucretius' text. Epicurus, the great master who haunts nearly every page of the poem, occurs by name only once.[45] Of course, his nominal absence differs from that of Memmius. At least in terms of his influence, Epicurus is almost omnipresent, and his name seems to hold, if anything, the status of a *nomen sacrum* ("holy name").[46] Although Epicureanism's relationship with divinity (and anything we could call sacred) is complex, it is well-known that Lucretius, at least on occasion, elevates his master to the status of a god, proclaiming most famously (at *DRN* 5.8) *deus ille fuit, deus* ("that man was a god, a god"). And it seems that Lucretius, at least within the Epicurean sect, may not have been alone. In his *Tusculan Disputations*, for example, Cicero remarks that he is "... often accustomed to wonder at the arrogance of some philosophers [*sc.* Epicureans], who admire the study of nature and exultingly pay thanks to its founder and chief, honoring him as a god" (1.21 [48]: ... *soleo saepe mirari nonnullorum insolentiam philosophorum, qui naturae cognitionem admirantur euisque inventori et principi gratias exultantes agunt eumque venerantur ut deum* ...).[47] While it may be too strong to accuse Lucretius of engaging in a version of negative theology,[48] it is striking that the only instance in which Epicurus' name appears is an instance that emphasizes the

45. Compare the comment at Bailey (ed.) 1947: 611: "[t]here is no doubt a feeling of taboo about mentioning the name of this *deus*..."

46. Such an observation, although not entirely familiar in Epicureanism, is hardly foreign to ancient philosophical discourse more generally. The name of Plato, for example, often takes on a divine, or quasi-divine, status in the history of Platonism. Lucretius' contemporary Cicero, for example, speaks of *deus ille noster Plato* ("our Plato, that god") at *ad Att.* 16.3 (a letter that dates to a time (mid-54 BCE) near the traditional date of composition assigned to the *De rerum natura*) and mentions Plato as *quasi quidam deus philosophorum* ("a certain god, so to speak, among philosophers") at *ND* 2.12.32.

47. Although it can only serve as a distant suggestion of Lucretius' own attitude toward Epicurus (and his use of the master's name), it is attested in the time of Epicurus that the Garden conducted itself as a religious society with its own festivals and rules of life. See Plutarch *Adv. Col.* 18 (1117D), where Epicurus seems to refer to the Garden as a "holy body" (ἱερὸν σῶμα), and the evidence in Clay 1986. A brief account may be found at Bendlin 2002: 9, with further secondary references in his n1; a fuller account is available in Koch 2005 (reviewed by Gaulin 2007).

48. To be clear, I use "negative theology" to refer to the linguistic practices commonly associated with speech about an ineffable deity. See Sells 1994 for a study of some of the linguistic practices of negative theology; his first chapter (14–33) on Plotinus is particularly relevant for students of ancient philosophy.

great master's mortality (as opposed, that is, to his divinity): *ipse Epicurus obit decurso lumine vitae* … (*DRN* 3.1042: "Epicurus himself perished, his life run out…").[49]

However we parse this one nominal intrusion into the poem, if we look more closely at Lucretius' direct references to Epicurus, we note that it is Epicurus' Greekness—rather than any other attribute of Lucretius' great master—that repeatedly surfaces in *De rerum natura*. At *DRN* 1.62–79, for example, where Epicurus first appears, championed as one who "traveled far beyond the flaming walls of the world" (1.72–73: … *extra/processit longe flammantia moenia mundi* …), he appears not by name but as *Graius homo* ("a Greek man"). Or again, when Lucretius praises his master's accomplishments at the beginning of his third book, he appears as a "you," "a Greek":

> *O tenebris tantis tam clarum extollere lumen*
> *qui primus potuisti inlustrans commoda vitae,*
> *te sequor,* ***o Graiae gentis decus*** *… (DRN 3.1–3)*

> *O, who from such great darkness first was able*
> *to raise up such shining light, illuminating the benefits of life,*
> *you I follow,* ***ornament of the Greek race*** *…*

Finally, in the sixth book, Lucretius makes a similar move, emphasizing the (Greek) location of Epicurus' philosophizing:[50]

> *Primae frugiparos fetus mortalibus aegris*
> *dididerunt quondam praeclaro nomine* ***Athenae***
> *et recreaverunt vitam legesque rogarunt,*
> *et primae dederunt solacia dulcia vitae,*

---

49. It is worth pointing out that at *DRN* 2.42 the manuscripts report *subsidiis magnis epicuri constabilitas*, although the nonsensical *epicuri* is rejected by all editors. Lucretius' practice in using or (more commonly) in not using Epicurus' name is difficult to locate against traditional Roman religious practice. It is familiar in Roman (but not only Roman) religious contexts to allow for the possibility of having mistaken or omitted a divinity's name; see Macrobius *Sat.* 3.9.10 (a *devotio*): … *sive vos quo alio nomine fas est nominare* … ("… or by whatever other name it is right to call you…"). Additional evidence may be found at Appel 1909: 75–79 and Wissowa 1912: 37n4. The suppression of a name—which seems to be occurring in the *DRN*—is in some ways more reminiscent of the Roman defense against *evocatio*, a ritual in which a tutelary deity was "called out" of a besieged city by name. The name of this deity was kept secret to protect the city. See Macrobius *Sat.* 3.8.9, Basanoff 1947.

50. While the three passages cited here could hardly represent the total effect of Epicureanism on the poem, they do nonetheless represent (together with the already mentioned opening to

*cum genuere virum tali cum corde repertum,*
*omnia veridico qui quondam ex ore profundit;*
*cuius et extincti propter divina reperta*
*divolgata vetus iam ad caelum gloria fertur. (DRN 6.1–8)*

***Athens** of famous name first brought forth*
*fruit-bearing crops for sick mortals in the past,*
*and regenerated life and brought laws,*
*and first gave sweet solace for life,*
*when it bore a man possessed of such wisdom,*
*who once poured forth everything from his truth-speaking mouth;*
*beacause of his divine discoveries, even with him dead,*
*his glory, long widespread, now is raised to the heavens.*

Lucretius' relationship to Greekness (including the Greek language), which must in some way figure in the repeated depiction of Epicurus as a Greek we see in these passages, has been much discussed in recent years.[51] While full treatment of this topic would exceed the bounds of the present discussion, it is worth noting briefly the opinions of those who have studied the question in greater detail. David Sedley, for example, remarks:

> Lucretius considers Greek culture artistically and philosophically superior, and yet at the same time deeply alien. He floods his poem with Greek words, but scrupulously avoids them in the course of doctrinal exposition.... [i]t is precisely by drawing attention to the culture divide between the Greek and the Roman, while making Epicurean philosophy nevertheless thoroughly at home in his own native language, that he proves to us its true universality.[52]

Joseph Farrell, writing only a few years later than Sedley, is not as willing to grant the superiority of Greek culture in Lucretius' eyes, commenting that "... there is no reason to assume that he [*sc.* Lucretius] sees the poem he has

---

the fifth book, where Epicurus is proclaimed a god) all the most significant direct references to the founder of the philosophy.

51. Two notable treatments that focus primarily upon Lucretius' relationship to the Greek language are Sedley 1998: 35–61 and Farrell 2001: 28–51. Their bibliographies collectively cover most extant literature on the question, particularly Lucretius' treatment and translation of Greek technical terminology.

52. Sedley 1998: 58–59.

labored to write as second best." Instead, "... Epicurus, forced to contend with the luxuriance of the Greek tongue, could not find a form adequate to his message; this task was left to his greatest disciple, who was also one of the greatest masters of Latin speech."[53]

Wherever one falls on the issue of Lucretius' perception of Greek cultural superiority, what is most striking (and at least implicitly agreed upon by both Sedley and Farrell) is the mere perception of and emphasis upon cultural difference, the sense—as Sedley phrases it—that Greek culture is "deeply alien."[54] This point may appear, in some ways, rather obvious, but there is an Epicurean (and Roman) sentiment behind the emphasis. If we return to the passage cited earlier from the sixth book of Lucretius' poem, we see a certain displacement, at least on the level of the name, of Epicurus by Athens, the place that "bore" (*genuere*) him. The articulation of the period, with its initial hyperbaton and repeated *primae* ("first"), creates a constant focus upon the grammatical subject, Athens, rendering Epicurus, though certainly not an afterthought, a product of the city in the sequential movement of the sentence.[55] This displacement, an effect that emphasizes both Epicurus' Greekness and Athens as the place of his philosophizing, makes a certain sense within an Epicurean framework. If we recall not only the Epicurean origins of language presented in the *Letter to Herodotus* (75–76)[56] but also Lucretius' comment on *res gestae* ("deeds; history") articulated in the first book of his poem—that "you could deservedly call them (*sc. res gestae*) *eventa* ["accidents"] of the body and of the place in which each occurred" (*DRN* 1.481–82: *merito possis eventa vocare/ corporis atque loci, res in quo quaeque gerantur*), then the focus upon

53. Farrell 2001: 51.

54. To some extent, the difference in opinion between Sedley and Farrell may be traced to the projects of their respective works. Sedley treats the Greek philosophical tradition received and transformed by Lucretius, while Farrell writes on various myths and presumptions about the Latin language, reinterpreting, in the case of Lucretius, his well-known assertion of the "poverty" of the Latin tongue (*DRN* 1.139, 1.832, 3.260) as something other than a linguistic inferiority complex.

55. This passage is reminiscent of *DRN* 1.714–33, where Lucretius praises not only Empedocles but also—and in fact much more extensively—his home island of Sicily for having "carried" (*gessit*) him. On this passage, see Farrell 2001: 43–45, who notes (44) how "Lucretius' respect for Empedocles ... is expressed ... through *geography* and language..." (Emphasis added.)

56. "... names (τὰ ὀνόματα) did not arise originally by coinage (θέσει), but the very natures of men experienced particular impressions and grasped particular images; and they expelled in a particular way air sent by each of the impressions and images, so that there would be difference among the races *across their locales* (ἡ παρὰ τοὺς τόπους τῶν ἐθνῶν διαφορά)." (Epicurus *Letter to Herodotus* 76)

Athens (and upon Epicurus' Greekness) seems almost expected. Epicurus' accomplishments are *eventa*, accidents, of Athens, the place where his philosophizing transpired.[57]

If this analysis is correct, then Epicurus' nominal displacement is mirrored by, if not a material displacement, a material contextualizing that calls into question a simple deictic understanding of the proper noun. If, materially speaking, Epicurus is not a purely autonomous agent but a product of Greece, a product of Athens, then calling him *Graius homo* rather than *Epicurus*—even as it fails to identify Epicurus uniquely—nonetheless captures something of the material situation suggested by Epicurean doctrine. Nothing comes from nothing, not even the father of Epicurean doctrine himself, and the problem of how names accurately represent this fact is one that troubles an easy notion of individual, "proper" reference. Names may perhaps pick out individuals, but this may be a misrepresentation when those individuals are understood as not entirely individual.

To put this point more formally, the trope through which Epicurus' name is displaced in *De rerum natura* is known as antonomasia, "which," as Quintilian says, "puts something in place of a proper name" (8.6.29: ... *quae aliquid pro nomine ponit* ...).[58] The rhetorical handbook addressed to Herennius explains the trope somewhat more fully, using the Latin term *pronominatio* in place of antonomasia:

> **Pronominatio** est, quae sicuti cognomine quodam extraneo demonstrat id, quod suo nomine non potest appellari; ut si quis, cum loquatur de Gracchis: "at non **Africani nepotes**," inquiet, "istiusmodi fuerunt." item si quis de adversario cum dicat: "videte nunc," inquit, "iudices, quemadmodum me **Plagioxiphus iste** tractarit." hoc pacto non inornate poterimus, et in laudando et in laedando, in corpore aut animo aut extraneis rebus dicere *sic*, uti cognomen quod pro certo nomine collocemus. (*Ad Herennium* 4.42)
>
> *Pronominatio* (Gk. ἀντονομασία) is a trope that—just as with a certain extra *cognomen*—indicates that which cannot be designated by its own proper name. For example, someone, when he speaks concerning the

57. Moreover, as Lucretius hints with the words *et extincti* ("even with him dead"), there seems to be some concern how, materially speaking, we understand the existence of Epicurus' teaching and renown now, in the absence of his physical presence. Athens, of course, remains; and we may perhaps say that Epicurus' fame and philosophy are accidents of this place.

58. On antonomasia, see Lausberg 1990: 300–302 (§§ 580–81). For an insightful, theoretical discussion of this trope (primarily focusing on the work of Leibniz), see Fenves 2001: 13–79; 215–19.

> Gracchi: "the **grandsons of Africanus** were not of this sort," he will say. In a similar fashion, someone, when he speaks concerning his adversary: "see now, judges," he says, "how **this Plagioxiphus** has regarded me."[59] In this manner, we shall be able, both in praising and in blaming, to speak not without ornament, on the body or the soul or external matters in the same fashion as a *cognomen* that we confer instead of a certain (proper) name.

Taken together, these two definitions point up not only the way in which antonomasia acts substitutively—calling Epicurus *Graius homo* instead of employing his name, for example—but also how these substitutions replicate the function of proper nouns. The author of the *Ad Herennium* in particular emphasizes, re-iterating the point twice, that *pronominatio* functions in the same fashion as giving an additional *cognomen*, an observation that serves to reduce the distance between common and proper nouns. We have already seen numerous instances within the Roman tradition—names such as Rufus, Agrippa, or Cordus—where the distance between a proper noun and an adjective seems slight, defined, at least in a significant way, by the editorial practice of capitalization.

This point, that antonomasia both replaces proper nouns and verges on creating new proper nouns, is not only reminiscent of the fragment of Epicurus preserved by the *Theaetetus*-commentator (where common and proper nouns are rendered indistinguishable) but also provides the nominal analogue of the material interaction that at once produces a unique Epicurus and yet undermines his status as a singular, autonomous agent. (In some ways, this predicament is perfectly captured by the term *pronominatio*, which always names in place of [*pro-*] a proper name but never is genuinely proper itself. This process, if projected backward, creates an infinite regress, where we must always presume a prior proper name and nonetheless wonder how any name ever became "proper" in the first place.) Indeed, to specify and develop this parallel between the trope antonomasia and a materialist conception of causation, I would like to speak of a certain atomology, by which I mean not what is often meant by this term—that as Lucretius himself points out there is a similarity between the *elementa*, or atoms, that structure the world and the *elementa*, or letters, that structure his poem[60]—but instead that words,

59. "Plagioxiphus" is a fictitious name meaning "Bruiser." See Lewis-Short s.v.

60. It has long been recognized that a strong current of etymological wordplay runs through much early and classical Latin poetry. Studies such as Ahl 1985 and, in a somewhat more cautious vein, O'Hara 1996 have investigated both the ancient theoretical underpinnings of

specifically the proper names that identify (and seemingly point to) individuals, interact like—and manifest properties of—atoms.

### *(c) Atoms and Anonymous Lovers*

The point already has been made that there is extensive overlap in the manner in which atoms and human agents are characterized. Atoms congregate and show emotions like humans, while humans band together and disperse like atoms. The present examination extends this realization by pointing to the similarities between the proper names that characterize human agents and the depiction of atoms in Lucretius' text. These similarities, the intermingling between atoms and names, perhaps push us along a path toward a materialism of language: as the material origins of language in Epicurean thinking condition difference from place to place, the material analogy between name and atom allows us, as we shall see, to think the anonymity and disappearance of names that we have already seen in our study of Lucretius' use of the names of Epicurus and Memmius within *De rerum natura*.

To be more specific, I would like to read the conclusion to Lucretius' fourth book, the so-called diatribe against love, as a window upon one set of

---

such wordplay and have catalogued numerous instances of it, especially in Virgil (O'Hara) and Ovid (Ahl). Somewhat similar, though more interested in semantics than phonetic wordplay, is Paschalis 1997 (on Virgil's *Aeneid*). In Lucretian studies, a related vein of criticism has grown up under the rubric of atomology, a term first coined by Friedländer 1941. Atomology, as traditionally understood, points to an analogy, or perhaps something more than an analogy, between letters and atoms first remarked by Lucretius himself, most notably at *DRN* 1.907:

> *iamne vides igitur, paulo quod diximus ante,*
> *permagni referre eadem primordia saepe*
> *cum quibus et quali positura contineantur*
> *et quos inter se dent motus accipiantque,*
> *atque eadem paulo inter se mutata creare*
> *ignes et lignum? quo pacto verba quoque ipsa*
> *inter se paulo mutatis sunt elementis,*
> *cum ligna atque ignes distincta voce notemus. (DRN 1.907–14)*
>
> *Do you now see—what we said a bit before—*
> *that it often matters quite a bit with which things*
> *and with which arrangement the same elements are*
> *held together and which movements they create and incur*
> *among themselves and that the same elements, a bit rearranged*
> *among themselves, make fire (ignes) and wood (lignum)?*
> *In which way, too, are the words themselves when their*
> *letters have been changed a bit among themselves,*
> *although we note wood (ligna) and fire (ignes) with distinct words.*

There has long been debate about the scope and nature of the comparison suggested by these lines. Relevant works (an abbreviated list) include Snyder 1980, Dionigi 1988, Armstrong 1995; and (against Snyder, Dionigi, and atomology more generally) West 1982 and 1991. Some of my own thoughts on this issue appear in Shearin (forthcoming).

Lucretian naming practices. However strange it may sound to read a discourse on love as writing about names, the final sections of Lucretius' fourth book not only permit such a reading but invite it, at once extensively employing names and explicitly contemplating—in ways we shall take care to examine—naming practices. One of the most explicit (and historically most puzzling) statements on names runs as follows:

> *Haec Venus est nobis; hinc autemst nomen Amoris;*
> *hinc illaec primum Veneris dulcedinis in cor*
> *stillavit gutta, et successit frigida cura.*
> *nam si abest quod ames, praesto simulacra tamen sunt*
> *illius, et nomen dulce obversatur ad auris. (DRN 4.1058–62)*
>
> *This is "Venus" for us; from here arises the name "Amor";*
> *from here first that drop of the sweetness of "Venus" fell*
> *into the heart, and icy care came, too;*
> *for if what you love is absent, nonetheless visions* (simulacra) *of it*
> *are present, and its sweet name flitters about our ears.*

The latter part of this statement articulates a now familiar sentiment about *simulacra* that links them with names, or at least with names that function in a deictic sense: the *simulacrum* and the "sweet name" (*nomen dulce*), themselves quite present, both point to an absent lover. Intriguing as these lines are, the initial lines of the statement, more puzzling and ambiguous, have attracted greater scholarly attention. The words *Haec … hinc … hinc* ("This … from here … from here") form a sequence that, appropriately enough for the questions we have been investigating, relies upon deixis: each of these terms points, although it is unclear precisely where. Moreover, *nomen Amoris*—or *nomen amoris*, as a different editor may have it—is a puzzling phrase, permitting of several translations. The above rendering "the name 'Amor'" presents only one interpretation; others include "the name of [the god] Amor" and "the word 'amor.'"

To better grasp where these lines point as well as their possible meanings, it is necessary to read backward and consider the passage that immediately precedes the present lines. The above passage comes immediately on the heels of one of Lucretius' more famous descriptions, that of the wounded lover, an image that juxtaposes love and war, seemingly drawing upon (among other things) the traditions of Greek erotic epigram:[61]

61. For discussion of some parallels and antecedents to Lucretius' presentation here—particu larly parallels to the "wounded lover"—see Kenney 1970: 380–92 and Brown 1987: 191.

> *... omnes plerumque cadunt in vulnus, et illam*
> *emicat in partem sanguis unde icimur ictu,*
> *et si comminus est, hostem ruber occupat umor.*
> *sic igitur Veneris qui telis accipit ictus,*
> *sive puer membris muliebribus hunc iaculatur*
> *seu mulier toto iactans e corpore amorem,*
> *unde feritur eo tendit gestitque coire*
> *et iacere umorem in corpus de corpore ductum;*
> *namque voluptatem praesagit muta cupido. (DRN 4.1049–57)*
>
> *In general, all fall into a wound, and blood spurts*
> *out into that direction whence we are struck with a blow,*
> *and if the combat is hand-to-hand, red fluid covers the enemy.*
> *So, then, whoever accepts a blow from the weapons of Venus,*
> *whether a boy strikes him with his feminine limbs*
> *or a woman, tossing lust from her entire body,*
> *whence he is smitten, to that spot he extends and strives to go*
> *and to toss his gathered fluid from body to body;*
> *for dumb lust presages pleasure.*

Reading this entire passage together with the lines cited above has allowed Paul Friedländer to produce the following reading:

> Interpreters usually refer *haec* [*sc.* at 1058] to *voluptatem, hinc* to *cupido* of the preceding verse.[62] But *haec ... hinc ... hinc ...* refer to the whole preceding process and *nomen Amoris* is not Cupido but just "the name Amor." By *hinc est nomen Amoris* the poet points to the twice-repeated *umor* (1051, 1056), as a few lines later he will again put side by side *umorem—amore* (1065–6).[63]

This reading is, in some ways, underwritten by Friedländer's focus upon reading Lucretius aloud and paying attention to the ways in which the words and sounds in the poem resonate materially. Such a focus explains, in part, the attention to the nearly homophonic pair *amor-umor*. Yet a second issue raised, perhaps more prominently, by Friedländer's interpretation, an issue that will be of importance for the present discussion, is the question of scope of reference.

62. These interpreters include both of the major multi-volume commentaries on Lucretius from the first half of the twentieth century. See Ernout-Robin 1926: 2.284, Bailey 1947: 416–17.

63. Friedländer 1941: 18. See, too, Brown 1987: 201–2, who follows Friedländer.

This question is first posed in terms of the deictic sequence *haec ... hinc ... hinc*: does this sequence refer merely to the line (1057) immediately preceding the declaration that "This is Venus for us" (*Haec Venus est nobis*)?[64] Or does this sequence refer to "the whole preceding process," as Friedländer insists? Any answer to these questions depends, at least in part, upon whether we view each of the terms (*Haec, hinc, hinc*) as having a specific or a more generic scope.

A somewhat different variation on this same theme plays out with regard to the phrase *nomen Amoris*, which could, on an alternative interpretation, be written *nomen amoris*. The distinction between these two possibilities, although it may seem an editorial choice of trifling significance, again concerns the perceived scope of reference: does *umor* ("fluid")—in the statement of the text—underlie the name of a specific divinity, Amor ("Love"), or a general, common noun ("love")?[65] Indeed, this question is itself a variation on a question we considered with regard to the functioning of the trope antonomasia, which shuttles between proper nouns and other appellatives. "Antonomasia," as Heinrich Lausberg reminds us, "is synecdoche for personal names: to the *genus pro specie* ('genus for species') of synecdoche corresponds a *species pro individuo* ('species for individual') in antonomasia."[66] This action, described by Lausberg, of placing the "species for <the> individual" is, in a way, the entire problem of the *Amor/amor* distinction: *Amor* is an *individual* divinity that embodies and stands for a whole *species* of experiences. And while this problem, the problem of *species pro individuo*, may, at first glance, seem rather insignificant, it seems to recur continually and even to dominate Lucretius' diatribe against love.

A question analogous to that posed by the *Amor/amor* distinction is asked repeatedly throughout the concluding sections of the fourth book by the appearance of (the name) *Venus*. In the final 250 lines of the book, the name of Venus appears twenty-three times, demanding various creative translations from those who attempt to render the poem in English and thus constantly calling into question how one proper name can cover this range (that is, this species) of phenomena.[67] To take but one major, published translation, the

64. On this interpretation, as Ernout-Robin suggest (1926: 2.284), *haec* refers to *voluptas* ("pleasure"); *hinc* means *ex cupidine* ("from *cupido* ['lust']").

65. See Feeney 1998: 87–92, esp. 88, where he discusses how "any imposing abstract word can indeed look awfully like a divinity in the right context."

66. Lausberg 1990: 301: "Die Antonomasie ist eine Synekdoche für die Eigennamen: dem *genus pro specie* der Synekdoche entspricht in der Antonomasie eine *species pro individuo*."

67. The density of usage of *Venus* is, of course, not surprising, given the subject matter treated. Nonetheless, it is striking that *Venus* occurs only thirty-four (34) times total (in *DRN*), including the hymn introducing the first book.

Loeb edition of W. H. D. Rouse (revised by Martin Ferguson Smith[68]) offers the following renderings for *Venus*, as it is employed in its various forms at turns throughout the concluding lines of the fourth book: "Venus" (repeatedly[69]), "Venuses" (a plural, 1185), "wedlock" (1235), "sexual" (1248), "penis" (1200, 1270), "intercourse" (1276). Another recent edition—that of Robert D. Brown—translates *Venus* at every turn with "Venus,"[70] a practice that preserves some of the flavor of the original Latin but also produces innovative turns of phrase like "the Venus of the mounting males" (1200: *Venerem salientum*).[71] The review of these translations aims not so much to criticize one or the other effort at bringing Lucretius' poem into English, but to attempt to understand the phenomenon of designating (or trying to designate) a wide range of phenomena with a single proper name. The referential scope of *Venus* changes constantly throughout the final lines of Lucretius' fourth book, a fact that not only—as many scholars have realized—problematizes the relationship of these lines to the benevolent *Venus* invoked as an "ally" (1.24: *socia*) in the first book of the poem[72] but also calls into question the function of *Venus* as a proper noun.

There is, in Latin, no "common" version of the noun *Venus*—a hypothetical *venus*—that corresponds to *amor* in our earlier pair. *Venus* is thus left to play the role of both common and proper nouns, and—perhaps as a partial consequence—is passed around metonymically throughout the final lines of the poem, not entirely unlike the *umor* ("fluid") that, within a Lucretian framework, motivates sexual desire.[73] While in the scene that concludes Lucretius' fourth book, *Venus* begins as a recognizable agent who acts with weapons (1052: *telis*)—and indeed continues as an agent who offers pleasure (1085), teases lovers (1101), even enjoys propitiation (1157)—the term

68. I refer here to the 1992 edition of the Rouse-Smith work, which—while not claiming to be a third edition—is a revision of the 1982 edition (that already called itself a "revised second edition").

69. *DRN* 4.1052, 1058, 1059, 1071, 1073, 1084, 1101, 1107, 1113, 1128, 1148, 1157, 1172, 1204 (1205), 1215, 1223, 1278.

70. Making various accommodations for number and case usage, of course: for example, he renders the Latin plural *Veneres* (at 1185) with a plural "Venuses."

71. In Latin, Lucretius—at least judging from the surviving record—seems to pioneer this euphemistic usage of *Venus* (apparently meaning "penis"), although it is picked up by several later authors, e.g., Martial 1.46.2, 3.75.6; Juvenal 11.167. See Adams 1982: 57 and Brown 1987: 315 (*ad loc.*).

72. On the issue of reconciling Venus in Lucretius' proem with her depiction in the fourth book, see Brown 1987: 91–100.

73. See *DRN* 4.1063–65: ... *decet* ... / ... /*et iacere umorem conlectum in corpora quaeque*: "... it's fitting to toss the gathered fluid into any body whatsoever..."

becomes increasingly more material, referring first to women (1185: *Veneres*) who conceal the "behind-the-scenes activities of their lives" (1186: *postscaenia vitae*) and then (apparently) to the male genitalia (1200, 1270). It would no doubt be a mistake to read *Venus* simply and straightforwardly as a term that, over the course of a few hundred lines, descends from describing a quasi-divine[74] agent to referring to the materiality of bodily organs.[75] For one, Lucretius interrupts this descent, framing his discussion that began from the *telis Veneris* (1052: "weapons of Love") with a reference to the *sagittis Veneris* (1278: "arrows of Love"), an image that conjures up at least a fleeting sense of an arrow-wielding agent. And, while it is true that the sentiment expressed in Lucretius' diatribe is often understood as dislodging love (that is, Love) from a divine perch, the phenomenon played out with the name *Venus* seems better characterized by one of the adjectives Lucretius gives it himself. At line 1071 Lucretius speaks of a *volgivaga Venus* ("free-wandering love/sex"), a description usually taken to reflect the more or less atomic conception of love he advocates, but *volgivaga* could just as easily characterize the word *Venus* and its promiscuous wandering among a whole range of referents.

The atomic quality of this referential wandering should be stressed, not only because one of the primary metaphors for describing atomic motion, a martial one,[76] resonates well with descriptions we have seen already of "love" (e.g., the wounded lover falling into his wound) but also because

74. The divinity of Venus is difficult in an Epicurean context. A famous Epicurean *dictum* proclaims οὐδὲ θεόπεμπτον εἶναι τὸν ἔρωτα (Diogenes Laertius 10.118: "that love is not divinely sent"), but this statement appears to contrast with Lucretius' own invocation of Venus to begin his poem.

75. In his materialist depiction of Venus (emphasizing bodily and sexual aspects), Lucretius may have drawn inspiration from his forebear Ennius (cf. *DRN* 1.116–26), who depicts Venus as the founder of the art of prostitution in his *Euhemerus*. Ennius *Var.* 142–45 Vahlen [*Euhemerus*]: *Venus prima artem meretriciam instituit auctorque mulieribus in Cypro fuit, uti vulgo corpore quaestum facerent: quod idcirco imperavit, ne sola praeter alias mulieres inpudica et virorum adpetens videretur.* ("Venus first established the art of the prostitute and was an *auctor* for women in Cyprus, that they might commonly pursue gainful occupation with their bodies: which thing she commanded lest she alone—beyond other women—seem shameless and desirous of men.")

76. Above all, we should recall the description of atomic motion via the comparison of atoms to motes in a sunbeam. See *DRN* 2.116–122: *multa minuta modis multis per inane videbis/ corpora misceri radiorum lumine in ipso/ et* ***velut aeterno certamine proelia pugnas/ edere turmatim certantia*** *nec dare pausam,/ conciliis et discidiis exercita crebris;/ conicere ut possis ex hoc, primordia rerum/ quale sit in magno iactari semper inani.* ("You will see many tiny bodies variously mixed/ through the void in the very light of the sun's rays/ and, **as if with endless struggle, they produce,/ in attacking throng by throng, battles and fights**,/ and do not cease, occupied with constant assemblies and discord;/ so that from this you could guess of what sort are/ the constant tossings about in the great void.")

the substitution *species pro individuo* ("species for individual") continually exploited by Lucretius is itself, in many ways, atomic. *Venus* (the name), as we have seen, functions throughout the conclusion of the fourth book as a metonymy for a whole range of entities ("penis," "intercourse," etc.) connected with the sphere of love, but it also—in a somewhat different sense—functions as a frustrated antonomasia: *Venus*, the specific, individual (proper) name, is constantly describing a more general range (i.e., a *species*) of experiences, but without, of course, the substitution of expressly more general terms. *Venus*, that is, becomes *species pro individuo* in itself, no longer functioning (if it ever did) merely as a deictic marker for (the perhaps imagined *simulacrum*[77] of) a divine agent. And this function, *species pro individuo*, is a function of the atom itself.

One of the notable facts about atoms, although in combination they form the myriad wonders of the natural world, is that they have no secondary qualities.[78] Lucretius is adamant about this point. At *DRN* 1.830–920, for example, Lucretius combats Anaxagoras' theory of the *homoeomeria* of things, which, our poet reports, supposes that bones "arise from tiny, minute bones" (1.835–37: ... *e pauxillis atque minutis/ ossibus ... / ... gigni* ...), flesh from tiny flesh (836–37), blood from drops of blood (837–38), and so forth. He refutes this claim by giving examples where the elements that contribute to a larger structure seem undeniably different from that structure, that is, food (*cibus*) increases and nourishes the body (*corpus*), even as it is scarcely of the same nature (859–74). Lucretius continues on, eventually telescoping this line of argumentation to the following conclusion (916–20): if we suppose that the elements of matter (*corpora materiai*) are endowed with a nature similar to larger entities they form, then it will come about that (*fiet uti*) atoms shake with laughter and soak their cheeks with salt tears.

These arguments and examples suggest Lucretius' reluctance to allow atoms any secondary qualities, although they do not necessarily make the point entirely explicit. Elsewhere, after Lucretius has informed us not only of the existence of atoms but also of the complexities of their movement, he

77. It is difficult to know what material existence to give to the notion of Venus as a divinity, but as Centaurs and Scyllas are given the existence of combinations of images (that is, *simulacra*), it seems safe to presume that we may have a *simulacrum* of Venus as a divine agent, if nothing more. See *DRN* 4.732–48.

78. I use "secondary property" (the standard term) in recognition of the fact that atoms do, of course, have some properties, not least their status as indivisible *corpora* ("bodies") that move.

states more emphatically the absence of secondary properties. Atoms, for example, have no inherent color:

> *Nunc age dicta meo dulci quaesita labore*
> *percipe,* ***ne forte haec albis ex alba rearis***
> ***principiis esse, ante oculos quae candida cernis,***
> ***aut ea quae nigrant nigro de semine nata;***
> ***nive alium quemvis quae sunt imbuta colorem,***
> ***propterea gerere hunc credas, quod materiai***
> ***corpora consimili sint eius tincta colore.***
> ***nullus enim color est omnino materiai***
> ***corporibus, neque par rebus neque denique dispar.*** *(DRN 2.730–47)*
>
> *Come now, know my words, sought out by*
> *my sweet toil,* ***lest perhaps you think these white things,***
> ***which you perceive shining before your eyes, are from white atoms,***
> ***or these, which are black, arose from a black seed;***
> ***or which things have been imbued with some other color,***
> ***lest you believe that they do this because the atoms of matter***
> ***are tinged with a similar color; for there is no color at all for***
> ***the atoms of matter, neither like nor again unlike things themselves.***

Nor do they possess heat, carry sound, or emit scent:

> ***Sed ne forte putes solo spoliata colore***
> ***corpora prima manere, etiam secreta teporis***
> ***sunt ac figoris omnino calidique vaporis,***
> ***et sonitu sterila et suco ieiuna feruntur,***
> ***nec iaciunt ullum proprium de corpore odorem.*** *(DRN 2.842–64)*
>
> ***But—lest by chance you think first bodies abide,***
> ***deprived of color alone—they also are separated utterly***
> ***from warmth and cold and hot vapor,***
> ***and they are borne without sound and deprived of juice,***
> ***and they do not emit any scent of their own from their bodies.***

These examples do not enumerate (in an impossible, comic gesture) *all* the qualities that atoms lack, but one infers inductively the strange absence of defining features that characterizes atoms. The absence of secondary properties creates a situation in which atoms—even as they are the most basic, indivisible elements of matter—are not the most basic instantiations of all properties; that is, atoms, even as they underlie all things, cannot, taken individually, explain the diversity of those things.[79] Atoms are thus, not wholly unlike the term *Venus* (at least as we have studied it in Lucretius' lexicon), *species* and *individuum* in one: *individuum* (literally, "not divisible," like the Greek *atomos*[80]) insofar as they are basic and primary; *species* insofar as they are generic and common.[81]

This characterization of the atom—if we again take the atom as an analogue for the name—perhaps fits best with the characterization of *Venus* just presented, but it also resonates with the characterization of Lucretius' use of the names of Memmius and Epicurus presented earlier.[82] The names of Epicurus, the specific, individual founder of Lucretius' philosophical faith who is nominally displaced by his own ethnicity, and of Memmius, the addressee who is largely replaced by a vague series of second-person imperatives and pronouns (*te, tibi*, etc.), are both atomic in so far as they represent *loci* of tension between a

79. "Explain" here must be understood in a particular sense. Lucretius puts forth the notion that atoms can combine and produce—without divine intervention—qualities not inherent in atoms themselves. (See, e.g., *DRN* 2.973–90, on sensation.) In this sense, Lucretius feels that atoms do explain the material world. Yet the functioning of atoms, *how* they create and generate qualities beyond themselves, remains as mysterious as any reference to divine miracle.

80. It is worth noting in this context that Cicero uses the Latin term *individuum* to translate the Greek *atomos*. See, e.g., Cicero *De Fin.* 1.17 (*corpus individuum*).

81. See, e.g., *DRN* 1.894–96: . . . / *scire licet non esse in rebus res ita mixtas,/ verum* ***semina*** *multimodis inmixta latere/ multarum rerum in rebus* ***communia*** *debent.* (". . . we may know things are not thus [*sc.* as Anaxagoras thinks] mixed up in things, but **seeds, common** to many things, ought to lurk variously mixed into things.") The notion of common—indistinct and indistinguishable—seeds conjures up the generality characteristic of a species, whose unity is based on ignoring individual differences.

82. The connection I have been making here between atoms and names is largely an associative one. That is, I have suggested that, not unlike the way(s) in which the depiction of atoms in Lucretius' text bears a certain similarity to that of human agents, atoms and names are represented similarly. Yet, as David Sedley (1996: 87) makes clear, there may be a still deeper philosophical connection between atoms and names (which would perhaps confirm the connections I have noted on the level of figuration): "Plato, Aristotle and the Stoics are teleologists, who regard the whole as ontologically prior to the part: the part can only be fully understood by reference to its function within the whole. Epicurus by contrast is an atomist. He standardly treats parts as discrete items which, in coming together, generate larger complexes—be they atoms forming phenomenal bodies, or humans forming societies—but which in no sense have that as their pre-existing nature or function. Even bodily parts like hands and tongues came into being before any functions—including their communicative functions—were found for them. On this same anti-teleological model, Epicurus regards the central core of language as an original set of naturally uttered 'names' . . ., correlated to individual objects or contents of experience, and only at a later stage supplemented and inflected into a full-scale language."

specific, distinct individuality and a broader generality (of ethnicity or of readership). And this nominal tension (that is, a tension located within the name) between individuality and generality represents, we may suggest, Lucretius' response to the intimacy of Catullan deictic naming that served as one of our points of departure. Lucretius does not fully abandon the Catullan deictic gesture that conjures up a real personal agent but his pointing always seems somewhat vaguer, more general, and more anonymous than Catullan practice.

This observation is perhaps best captured by a striking passage within Lucretius' diatribe against love, a passage where Lucretius runs through a catalogue of names that lovers, deluded by love, may give the (in this case, female) objects of their affections:

> *atque alios alii inrident Veneremque suadent*
> *ut placent, quoniam foedo adflictentur amore,*
> *nec sua respiciunt miseri mala maxima saepe.*
> *nigra "Melichrus" est, inmunda et fetida "Acosmos,"*
> *caesia "Palladium," nervosa et lignea "Dorcas,"*
> *parvula pumilio, "Chariton Mia," "Tota Merum Sal,"*
> *magna atque inmanis "Cataplexis Plenaque Honoris."*
> *Balba loqui non quit—"traulizi"; muta "Pudens" est;*
> *at flagrans odiosa loquacula "Lampadium" fit;*
> *"Ischnon Eromenion" tum fit, cum vivere non quit*
> *prae macie; "Rhadine" verost iam mortua tussi;*
> *at tumida et mammosa "Ceres" est "ipsa ab Iaccho,"*
> *simula "Silena ac Saturast," labeosa "Philema."*
> *cetera de genere hoc longum est si dicere coner.* (*DRN* 4.1157–70)
>
> *And some lovers mock one another, and they urge that*
> *they propitiate Venus, since they suffer with shameful love,*
> *and, wretched, they do not look upon their own ills, often great*
> *ones.*
> *A swarthy woman is "Melichrus," a dirty, smelly one is "Acosmos,"*
> *blue-eyed is "Palladium," tough and muscular is "Dorcas,"*
> *itsy-bitsy dwarf is "Chariton Mia," "Tota Merum Sal,"*
> *great and savage is "Cataplexis Plenaque Honoris."*
> *Balba (the stutterer) who cannot speak—"traulizi"; the*
> *mute is "Pudens";*
> *and the hateful, feisty, talkative one becomes "Lampadium";*
> *then she becomes "Ischnon Eromenion," when she cannot live*

> *from her wasting away; truly she is "Rhadine," who is already*
> *dead with her cough; swollen and busty, she's "Ceres ipsa ab Iaccho,"*
> *snub-nosed is "Silena ac Satura," full-lipped "Philema."*
> *It's a lengthy task if I should try to name other things of this sort.*[83]

While there are no doubt many ways to parse these lines,[84] perhaps the most striking feature within the context of the present discussion is how difficult it is to assign a status to the names, or "names," listed here. Is "Dorcas" (i.e., "antelope" or "gazelle" in Greek) a name, a nickname, or a descriptor? Or is it perhaps several of these at once? The passage suggests that each of these names is a way of re-naming—and potentially misnaming—reality, but this fact certainly does not exclude these appellations from functioning as "proper" names in the manner of many Roman *cognomina*. A glance at R. D. Brown's recent commentary provides some sense of the range of scholarly views on the "names" in this passage:

> Pierre Boyancé ... has argued that the Greek words in Lucretius are all nicknames and should be printed with a capital letter. The fact that *Dorcas, Rhadine, Silene, Satura*, and *Philema* are attested as personal names undoubtedly illustrates the hypocoristic value of the words selected by Lucretius and may suggest that some were inspired by actual nicknames.... Many of the words, however, cannot be interpreted in this way (*acosmos, Chariton mia, cataplexis, traulizi*) and to impose a uniform status on the others is artificial. It is better to regard them as a miscellaneous collection of sentimental terms which differ in background and function. Some are literary, others colloquial; some laudatory descriptive terms, others casual endearments, which happen to overlap with the repertoire of formal names and quasi-permanent nicknames. None, however, needs to be interpreted as a fixed appellation.[85]

83. In the translation offered here, I have capitalized all the "names" in the list, less out of deep conviction that this choice is correct than out of need to make some editorial choice. As the discussion in the main text makes clear, not all scholars follow this practice.

84. One topic that demands fuller treatment than the present discussion can provide is the gendering of Lucretius' reading of love. Gordon 2002, which explores how Lucretius plays with a traditional, Roman male gaze, is the most extensive treatment of this topic to date.

85. Brown 1987: 281–82 (commentary *ad loc.*).

There are at least two competing, if not conflicting, impulses in this description. On the one hand, Brown resists the conclusion (a claim that he attributes to Boyancé[86]) of proclaiming all the "names" in Lucretius' catalogue "proper" (that is, editorially speaking, he resists the choice of capitalizing them), insisting instead that some of these terms simply "cannot be interpreted in this way." At a minimum, then, Brown shies away from understanding these names as proper deictic markers pointing to specific individuals. On the other, Brown notes what he calls the "hypocoristic value" of the terms within the list, observing that some of these terms were "inspired by actual nicknames." This gesture (that is, searching for the reality of the names in the passage), even if it does not mark a conflict with the first impulse, manifests a contrasting interest in the intimacy (and reality) created by names, in the ability of the names to conjure up (and point to) a genuine romantic context. In trying to assess the extent to which the "names" offered here could point to a possible historical "reality," Brown's critical stance thus resonates with the traditional interpretation of our first instances of deictic naming (e.g., searching for the reality of Catullus' "Lesbia").

But, as these two impulses within Brown's commentary begin to show, there is a clear distance between Lucretius and Catullus. Whereas Catullus' deictic naming creates a genuine (even if misleading) sense of presence and intimacy, Lucretius' list—in part because it is so obviously a list, a set of examples—fails to mimic this effect.[87] These names—*Dorcas, Rhadine*, and others—hover between proper and common nouns, conjuring up the terms of our earlier discussion of *Amor*: they are specific "names," but they are defined and motivated by an entire class of qualities (e.g., *Dorcas* could name, and its appearance in the catalogue depends upon its ability to name, all "wooden, muscular" objects of affection). Each of these names at best points awkwardly and indistinctly to an individual, an individual whose only known feature is, we infer, shared by an entire class.[88]

86. But Boyancé is not alone in his view. See Colin 1955: 866: "*Pour chacun de ces mots grecs, nous sommes en présence d'un surnom personnel, accepté avec orgueil par l'intéressée et tiré—comme* Βοῶπις *pour Clodia—d'un détail de ses avantages physiques et de sa beauté*..." (Emphasis original.) For references to Clodia as Βοῶπις, see, e.g.. Cicero *Ad Att.* 2.9.1. Colin suggests that the names are originally those of courtesans.

87. Nor, I suspect, does it try to mimic the Catullan effect.

88. The *general* scope of the names (i.e., the fact that these names fit several people and several contexts) within the list is suggested by many features, most prominently by *alios alii* (1157: "some ... one another") but also by the generic character of each of the attributes described.

Moreover, it is impossible not to note the obvious Greekness of the names. This fact has been variously interpreted, with some suggesting that it indicates Lucretius here follows a Greek original, while others simply maintain that Greek was (during the Roman Republic) "the natural vehicle for elegant or witty expression."[89] Whatever the case, within the present discussion, we cannot but be reminded of the way in which Epicurus' own Greekness (*Graius homo*) displaces his proper name in Lucretius' poem. Any author writing under the influence of a theory that language—language origins but also diversity among languages—is heavily conditioned by physical place cannot but be aware of a string of more than a dozen Greek "names" in his text. These names disguise "reality," mischaracterizing a set of less-than-charming attributes, perhaps in part because they are, quite physically, out of place.

This suggestion is somewhat speculative, but David Sedley has remarked, in a quite general fashion, "that the creation of a Greek context tends, in Lucretius' hands, to emphasise the remote and the exotic."[90] If correct, this reading would fall, in its own way, within the general pattern we have been tracking in this chapter, that is, how names in Lucretius become more remote, distant, and anonymous. While the terms of this list hardly disappear like Memmius' name, they nonetheless create a sense of remoteness through their almost euphemistic Greek cloaking of a fault more baldly stated in Latin. (Indeed, the contrast with the list of descriptors—actually a list of genuine Roman *cognomina*—at Horace *S.* 1.3.44–62 is instructive, as it makes clear how often Roman *cognomina* did *not* describe faults euphemistically.) Moreover, the general sense of substitution (the action, we might say, of antonomasia)—the Greek for the Roman, the common for the proper, the proper for the common—that surrounds discussion of the list continues the atomic qualities, or lack of qualities, we found in *Venus* (and *Epicurus* and *Memmius*). Despite or perhaps because of the lure of deictic naming,[91] it seems that names, like atoms, possess in practice no essential properties—beyond the simple fact that they are names.

89. Brown 1987: 281.

90. Sedley 1998: 51. Perhaps Greek words are remote and exotic in part because they are, physically, from elsewhere.

91. What I mean is that—as we saw with the comment preserved by the anonymous commentator on Plato's *Theaetetus*—deictic naming (an act of pure pointing) can in fact obscure distinctions between proper and common names. In other words, as deixis is an act (rather

## III. Conclusion: Anonymity and Popular Politics

Taking our point of departure from a selective consideration of Roman naming practices (and the connection of those practices with identity), we first developed in this chapter a notion of deictic naming, a naming that points (or attempts to point) to a given entity and often conjures up a sense of the reality of that entity. We then turned to examine the appearance of deictic naming within Epicureanism and studied its relationship to various instances of naming within Lucretius' *De rerum natura*. Through a study of the nominal disappearance of Memmius (or his metamorphosis into a "general" reader), the near nominal absence of Epicurus, and the deployment of *Venus* and other names within the conclusion to Lucretius' fourth book, we observed the ways in which various personal names in *De rerum natura*, even if they do not entirely fail to function in a deictic fashion, nonetheless become more general, more indefinite, whether by the substitution of other terms (Memmius, Epicurus), by referential wandering (Venus), or by still other means.

Having thus examined the ways in which a select group of names in Lucretius becomes more anonymous—vaguer, more general—we now turn briefly to the significance of this phenomenon. Why would Lucretius so strongly resist deictic naming with the three proper names most central to *De rerum natura* (his philosophical master, his addressee, and the "goddess" who opens his poem)? A recent author, who studies catalogues of names rather than naming effects, suggests that "Lucretius allows little textual ground for the treatment of proper names" for two reasons: "first, they are not important to him for the development of his subject; and second, his work contains a limited amount of mythological material, which is an abundant source of proper names."[92] The second of these reasons is certainly true: the Epicureans famously combated the fear and anxiety inspired by traditional religious and mythological explanations of the world; and scholars have much exerted themselves to justify even Lucretius' limited use of Venus.[93] The first reason, however, is less clearly on target. While Lucretius undoubtedly does use a limited number of proper names, as this chapter has endeavored to show, some of these—certainly the name of Epicurus—are of incredible, quasi-religious importance to him and

than an intrinsic feature of a word), it places an emphasis not upon the word itself but upon what the word does. It is nonetheless true that deixis butts up against concerns of presence and intimacy, a fact that often makes the specific name seem important.

92. Kyriakidis 2007: 87.

93. Gale 1994 and Gigandet 1998 provide recent scholarly takes on Lucretius' use of myth.

his poem. To say that proper names are "not important to the development of his subject" is also to ignore both the poetic and philosophical traditions handed down to Lucretius: Roman epic to his day was largely about particular historical figures, and philosophy—both earlier and later—often used proper names precisely as the poles around which speculative opinions revolved.[94]

Instead, we should consider the widespread absence of proper names and the tendency toward anonymity traced in this chapter as a powerful effect of Lucretius' doctrinal commitments. In tracking—among other evidence—the near nominal absence of Epicurus, the disappearance of Memmius, and the referential wandering of Venus, we have suggested that there are certain similarities between the material (absence of) qualities characteristic of atoms and the referential qualities of certain names. I would like to suggest in conclusion that there is a certain popular, everyman quality to the atomic tendency toward nominal anonymity, which we have been following. This suggestion is, on one level, merely to extend the comparison between atoms and names, since there is an ancient tradition of the democratic, egalitarian "community of atoms."[95] In other words, by rendering his poem less particular, less clearly stamped by the detail of deictic naming, Lucretius packages his salvific message in terms that are more broadly applicable and more generally accessible to even a moderately educated audience.[96] It is not merely names, after all, that are presented in general terms in *De rerum natura*; figures like the ploughman (*arator: DRN* 2.1164), who toils in vain at the conclusion to the second book, are depicted so sparsely that he could be many a Roman male.

---

94. For epic, if one thinks of Naevius and Cicero (e.g., *De consulatu suo*) in addition to Ennius, the Roman genre of epic seems to have been primarily historical (and about *named* historical agents) prior to Lucretius and Virgil; for philosophy, one need only look to Lucretius' contemporaries Cicero (e.g., *De natura deorum*, where we find Brutus, Protagoras, Diagoras Melius, and Theodorus Cyrenaicus already in the first chapter of Book 1) and Philodemus (e.g., his *Peri poematon*, where almost all of his opinions are developed agonistically with other named philosophers).

95. Cf., e.g., the remarks—part sarcastic, part serious—of Dionysius of Alexandria:

> θαυμαστὴ γε τῶν ἀτόμων ἡ δημοκρατία, δεξιουμένων τε ἀλλήλας τῶν φίλων καὶ περιπλεκομένων, εἰς μίαν τε κατασκηνοῦν συνοικίαν ἐπειγομένων … [Dionysius of Alexandria (apud Eusebium *PE* 14.25 776D)]
>
> Quite wondrous is the democracy of the atoms: they greet one another as friends, and they embrace, hastening to encamp together in a single settlement …

This statement is no doubt inspired, at some level, by the material equality of atoms, an equality underwritten, at least in part, by their lack of secondary qualities.

96. We may consider Epicurus' own demand for linguistic clarity (*sapheneia*; cf. Diogenes Laertius 10.13) to have a similar popularizing force.

Further, the popularization generated by Lucretius' nominal practice has other possible resonances among his Roman literary forebears, particularly in Cato the Elder's *Origines*. It is reported that *Origines* were written without giving credit to the leaders of the Punic wars:

> Senex historias scribere instituit. Earum sunt libri septem. Primus continet res gestas regum populi Romani, secundus et tertius unde quaeque civitas orta sit Italica, ob quam rem omnes Origines videtur appellasse. In quarto autem bellum Poenicum est primum, in quinto secundum. Atque haec omnia capitulatim sunt dicta. Reliquaque bella pari modo persecutus est usque ad praeturam Servii Galbae, qui deripuit Lusitanos: **atque horum bellorum duces non nominavit, sed sine nominibus res notavit**. (Cornelius Nepos *Cato* 3.3–4)[97]
>
> As an old man, he began to write *historiae*. There are seven books of these. The first book contains the deeds of the kings of the Roman people; the second and third books treat whence each Italic state arose, for which reason, it seems, the whole work was called *Origines* ("Origins"). In the fourth book, moreover, appears the first Punic war; in the fifth, the second Punic conflict. And these things were all told under headings (*capitulatim*).[98] He also, in like fashion, gave an account of the remaining wars up until the praetorship of Servius Galba, who plundered Lusitania: **he did not name the leaders (*duces*) for these wars, but he recounted things (*res*) without names**.

There are no doubt many possible explanations for Cato's pointed omission of what must have been well-known names, but one suggestion that has been voiced within recent scholarship on the *Origines* corresponds in certain ways to the line of explanation we have pursued in this conclusion:

> The concentration upon the history of a people, specifically upon the life of the *populus*, shifts the focus within the work. When the people itself becomes the primary actor, whence its origin, its birth, and its assembly become of central interest. And exactly this may be seen in Cato's *Origines*.[99]

97. Compare the similar description at Pliny *NH* 8.11.

98. *Capitulatim* ("under headings") is often taken to indicate a contrast with annalistic (year-by-year) narrative. Cf. Gotter 2003: 116.

99. Gotter 2003: 126: "Die Konzentration auf die Geschichte eines Volkes, gewissermaßen auf den Bios des *populus*, verschiebt die Gewichtung innerhalb des Werkes. Wenn das Volk selbst zum maßgeblichen Akteur wird, rücken seine Herkunft, seine Genese und seine

Ulrich Gotter's description here indicates one reason why Cato may have suppressed the names of leaders from recent wars: his interest fell upon the species (that is, the *populus*) rather than the individual.

Any thematic connection between Cato and Lucretius may be entirely coincidental, yet the fact that Cato suppresses the names of the primary agents in his tale of origins provides a literary precedent, if not for all Lucretian naming practices, at least for the virtual omission of Epicurus' name from *De rerum natura*. The *Origines* may seem generically distant from Lucretius' poem, but Lucretius certainly drew upon earlier historical epic (cf. *DRN* 1.117, 121), so it is not much of a leap to suppose some familiarity with the tropes of earlier historical writing. While there is no doubt more to be said on the connection between Cato and Lucretius, for the present purpose it suffices to observe the popular focus of Cato's anonymous narrative, or—more correctly—the popular focus of his narrative marked, we could say, by the figure of antonomasia, where Romanness seems to take the place of any specific, individual identity. We need not urge that Lucretius' focus was likewise on Romanness (or the "atomic" Roman people), but it has long been noted that Lucretius' natural world has a Roman coloring. Over seventy-five years ago, Hugh Sykes Davies pointed out how Roman the machinery of the world becomes in Lucretius' hands.[100] If the observations of this chapter are broadly correct and names are "atomized" through anonymity, then this process may indeed lay the groundwork for the depiction of the *populus Romanus* not only in benignly general terms but even as a faceless mass. This, then, is a double-edged populism: vaguely depicted, general agents are no doubt more equal, but a people without qualities is scarcely a people at all.

---

Zusammensetzung ins Zentrum des Interesses. Und genau das lässt sich in Catos *Origines* beobachten."

100. Davies 1931–32: 37–8: "'Coetus,' 'congressus,' 'foedus,' 'imperium,' 'indicium,' 'reddo,' 'usurpo,' all these are political and legal expressions that occur more or less frequently in Lucretian metaphors.... It is just possible that Lucretius may have attempted to represent the machinery of the universe—the 'machina mundi'—by symbols drawn from the legal and political machinery of the Republic. Such an idea is at least very typically Roman; they felt profoundly, throughout their history, that the Roman system of government had the same validity, the same unquestionable sanction that they found in the realm of Nature; sometimes it seems to them that their system was the direct outcome of the working of the natural world, that the Roman 'Imperium' was no more to be questioned than the heavens, its operation no more than the mutations of the seasons."

# 4

# *Catachrestic Origins*

## SPEECH ACTS AND THE POLITICS OF THE PERFORMATIVE

*Je devais à la fois marquer que j'étais une unité parmi d'autres, mais une unité impossible à chiffrer, perpétuellement excitée par sa propre fin …*[1]

WE RETURN NOW to the beginning—to the origins of language, that is, and to their role within Lucretius' anthropology (and the subsequent development of society) in the fifth book of *De rerum natura*. The preceding investigation, motivated initially by the coincidence of aspects of Epicurean thinking on language with modern discussions of speech acts, has butted up, at various turns, against political and societal elements in Lucretius' poem. Discussion in the second chapter suggested that Lucretius' figuration of the natural world painted him in the (occasional, structural) role of a fetial priest, a maker of international treaties and a spokesperson for the Roman people. In the third, atoms symbolized the anonymity observed in Lucretius' use of certain proper names, an anonymity that bears similarity to Cato's narrative focus on the people (*populus*) in his *Origines*. Both discussions relied for their conclusions on the conflation of descriptive language applied to humans with that applied to atoms. The present chapter studies a similar—and similarly political—strain in Lucretius' account of the origins of language and language's subsequent role in the development of human civilization.

To be more explicit, this chapter aims to locate a certain type of politics that links two persistent themes of discussion in earlier chapters: first, the significance of speech acts in and for the *De rerum natura* and, second,

1. Philippe Sollers, *Nombres* 2.10: "At the same time, I had to indicate that I was a unity among others, but a unity impossible to decipher, constantly excited by its own proper end …"

the figurative conflation of humans and atoms. This politics, which may be called—at least in a limited sense—"democratic,"[2] is perhaps best understood as the politics of the *unus quisque* ("each one").[3] *Unus quisque* here, as at various turns within Lucretius' text, designates an indefinite, generic individual, an individual that is at once specific and distinct and nonetheless typical of a class.[4] It occupies a position crossed by two competing forces. On the one hand, within Lucretius' anthropology (as elsewhere in the *De rerum natura*), there is a tendency to generalize, a tendency to speak generally and generically of the *genus humanum* ("human race"). On the other, there is a need to account specifically for the individual details of human experience, to work out even generic traits on an individual level.[5] The chapter aims to expand and complicate this bare description of the tensions that swirl about *unus quisque*, particularly (though not exclusively) through attention to the individual's relationship to language within Lucretius's anthropology. Discussion begins in the following section with questions of number and identity, two issues clearly at play in any definition of *unus quisque* (a *one* that is at the same time an *each* or *every*) and, as we shall see, central to subsequent consideration of the now familiar medium of the speech act, at least as that medium appears within Lucretius' anthropology.

Later sections of the chapter elaborate the "catachrestic potential" inscribed by Lucretius within the primal scene of language, the originary

---

2. The term democratic has been used, somewhat loosely, at various turns throughout this study. The previously cited remarks of Dionysius of Alexandria (ap. Euseb. *PE* 14.25 776D) show that democracy was connected explicitly with atomism in antiquity: θαυμαστή γε τῶν ἀτόμων ἡ δημοκρατία, δεξιουμένων τε ἀλλήλας τῶν φίλων καὶ περιπλεκομένων, εἰς μίαν τε κατασκηνοῦν συνοικίαν ἐπειγομένων ... ("Quite wondrous is the democracy of the atoms: they greet one another as friends, and they embrace, hastening to encamp together in a single settlement ...").

3. Other studies have recognized the significance of *unus quisque* for Lucretius' poem, but they analyze the phrase to different effect. See Segal 1990: 38, 40, 42, whose interest lies in Lucretius' use of contrasting perspectives. He describes how the Epicurean poet writes occasionally from the perspective of "far-off time" (38), occasionally from the "individualizing" (40) perspective of the *unus quisque*. The emphasis of the present chapter instead falls upon clarifying and elaborating the perspective of the *unus quisque*, which is, although individual (as Segal asserts), nonetheless also general and generic. For other significant treatments of Lucretian political thinking, see Nichols 1976, Fowler 1989, Schiesaro 2007a and b, and Kennedy 2013b.

4. Readers will immediately recognize the similarities between the argument of this chapter and the preceding one. In many ways, the "anonymity" of personal names described previously prefigures the politics described in this chapter.

5. The terms of this discussion owe much to De Lacy 1957 and Gale 2004b: 53–54. Both contrast the mechanical process of the material world with our own individual, human investment in that world.

moment of human linguistic discovery. Catachrestic potential here names the ability of language, in the hands of its individual users, not only to satisfy a need for communication or to overcome the poverty of linguistic expression (a poverty famously bemoaned, at various turns, by Lucretius) but also to misrepresent and, importantly for Epicurean doctrine, spread misconception. Discussion links this potential to Lucretius' anti-teleological worldview as well as to the "democratic" politics of the *unus quisque*. Closing remarks briefly relate the linguistic observations made in the chapter to (reported) linguistic practices within the Epicurean school.

## *I. Mass Counting: One and Many*

Structural and, more particularly, numerical relations are important to the articulation of any politics. It would, in fact, not be a gross exaggeration to assert that traditional political constitutions are understood (and commonly represented) as numerical configurations: one ruling over many (monarchy), few ruling over many (oligarchy), or many ruling jointly (democracy). While Lucretius nowhere explicitly endorses a given political constitution, texts already considered in this study read his atomism—and the faceless, material equality of atoms—as democratic, a conclusion no doubt based in part on atoms' apparent equality before the laws (or, as Lucretius more often says, the "compacts" [*foedera*]) of nature.[6] Relying upon such analyses, one could search for the ways in which Lucretius' text confirms or challenges (or perhaps: confirms and challenges) a democratic reading. To a certain extent, later sections of this chapter will undertake such a project. The current section lingers, however, on a preliminary, though related, consideration: how does Lucretius, implicitly or explicitly, define (human) individuals and their relationship to larger groups?

This question is broad in scope, and any response to it, particularly one as brief as the current one, must remain provisional and limited. Nevertheless,

6. Scipio Africanus suggests in a passage of Cicero's *De re publica* (*DRP* 1.53) that "equality before the law" (*aequabilitas iuris*)—presumably including natural laws—characterizes a "democracy" (*liberi populi*). Charles Segal (Segal 1990: 101) speaks, moreover, of "the Epicurean principle of ἰσονομία: life and creation are always counterbalancing death and decay, and vice versa." For those concerned with avoiding any terminological anachronism, though, democracy and democratic are difficult terms to apply to Lucretius (and his contemporaries) without qualification. To judge based upon Cicero, at the time *De rerum natura* was written there was not a simple Latin equivalent of the Greek term δημοκρατία. See *DRP* 1.42, where Cicero speaks of a *civitas popularis* ("popular state") and the commentary of Zetzel *ad loc.* (Zetzel 1995: 131).

the evidence that follows will prepare later consideration of how individuals and groups function within Lucretius' anthropology. In the ensuing discussion, I examine how Lucretius argues against the uniqueness of material compounds, while nonetheless maintaining the uniqueness of humans beyond their strict material composition. The analysis suggests that—even at the level of the human individual—there persists a tension (one we shall see replayed later) between, on the one hand, the generality underwritten by material constancy (that is, the consistent nature of the material components of individuals) and, on the other, individual specificity (and, therefore, difference), above all the specificity and distinctiveness of individual experience. The analysis will then turn to an examination of the boundary stone (*terminus*), which serves as a defining limit for various classes of individuals within the Lucretian world, a cultural marker of fairness and equality in a broad range of physical (and political) contexts.

## (i) One(s)

Lucretius' second book, containing his fullest descriptions of atomic motion and (perhaps most famously) the *clinamen* ("<atomic> swerve"), is one of his most materialist moments, insofar as he is focused on elucidating the mechanical functioning of the natural world at a microscopic level. Yet it is also a book that provides great insight into his characterization of human individuals, chiefly because the conclusion to the book appears to make one of the strongest statements against granting material individuals any uniqueness whatsoever. That conclusion evolves out of a series of arguments aimed at drawing a firm boundary between atoms and sentient beings. Lucretius first (*DRN* 2.865–990 *passim*) counters the notion (which he considers absurd) that sentient beings must be comprised of smaller sentient parts. The principle that any human (that is, macro-level) property (e.g., sentience) must be underwritten by the presence of the same property at the atomic level would entail, among other things, that atoms would laugh (976), cry (977), and investigate their own first-beginnings (979). The expectation that sentience exists for atoms themselves is thus, on such a view, ill-founded.

Despite this firm separation of human qualities from atomic ones, atoms do carry consequences for humans, as for other larger, atomic compounds. Most notably, since atomic movements are repeated constantly across boundless space, one ought not, Lucretius suggests, presume that material compounds are unique. Instead, material compounds recur, including our world and everything in it:

*nullo iam pacto veri simile esse putandumst,*
*undique cum versum spatium vacet infinitum*
*seminaque innumero numero summaque profunda*
*multimodis volitent aeterno percita motu,*
*hunc unum terrarum orbem caelumque creatum,*
*nil agere illa foris tot corpora materiai;*
*cum praesertim hic sit natura factus, et ipsa*
*sponte sua forte offensando semina rerum*
*multimodis temere incassum frustraque coacta*
*tandem coluerunt ea quae coniecta repente*
*magnarum rerum fierent exordia semper,*
*terrai maris et caeli generisque animantum.*
*quare etiam atque etiam talis fateare necesse est*
*esse alios alibi congressus materiai,*
*qualis hic est, avido complexu quem tenet aether. (DRN 2.1052–66)*

*In no way ought it be thought likely—*
*since an infinite space lies open in all directions*
*and atoms, countless in number, flit about in many ways—*
*in the greatest depths, continuously stirred up in eternal movement*
*that this world and heavens have arisen uniquely,*
*and that those many material bodies do nothing outside,*
*especially since this world arose naturally, and of their own accord*
*the very seeds of things, driven vainly by chance scraping,*
*variously, randomly, have cultivated things, which—when composed—*
*always become straightaway the beginnings for great things,*
*for the earth, the sea, the heavens, and the race of animate creatures.*
*Wherefore, again and again you must admit that*
*there exist other material assemblages elsewhere,*
*like this one here, which the sky holds in eager embrace.*

Important for the present discussion is the fact that, at least in this passage (and in its continuation), Lucretius lumps humans together with other material compounds. Animals of all kinds are not unique. They, including "the twin offshoot" (i.e., the two sexes) of the human race, are neither born nor grow uniquely:

*Huc accedit ut in summa res nulla sit una,*
*unica quae gignatur et unica solaque crescat,*

*quin aliquoiu' siet saecli permultaque eodem*
*sint* ***genere****. in primis animalibus inice mentem:*
*invenies sic montivagum* ***genus*** *esse ferarum,*
*sic hominum geminam prolem, sic denique mutas*
*squamigerum pecudes et corpora cuncta volantum.*
*quapropter caelum simili ratione fatendumst*
*terramque et solem lunam mare, cetera quae sunt,*
*non esse unica, sed numero magis innumerali,*
*quandoquidem vitae depactus terminus alte*
*tam manet haec, et tam nativo corpore constant,*
*quam* ***genus*** *omne quod hic* ***generatim****st rebus abundans.*[7] *(DRN 2.1077–89)*

*Moreover, in the sum (sc. of matter) there is no one thing*
*that is born uniquely and uniquely grows alone,*
*but for any age there are very many things*
*of the same* ***sort****. First of all, think of animals:*
*you will find that the* ***class*** *of mountain-wandering wild animals is so;*
*so the twin offshoot that makes up the race of men;*
*so, finally, mute fish and all birds.*
*For which reason, it must be admitted, following a similar account,*
*that earth and sun, moon, sea, all things that exist—*
*they are not unique but more than innumerable in number,*
*since the boundary stone of life awaits them, driven as deeply,*
*and they consist in body as mortal,*
*as* ***for any race, in its various kinds****, that here abounds in things.*

Humans here are treated as generic items, as the repeated use of *genus* (and *generatim*, a plausible emendation) stresses.[8] Individuals are interchangeable with other material entities of the same class, all alike before "the boundary stone of life" (*vitae ... terminus*) [that is, death]. The architecture of the passage stresses the point: its lines unfold through an accumulative process of assimilation. With each *sic* ("so, thus") or *et* ("and"), another group is subsumed under the same governing principle. Such treatment does not in any way render the numerically singular individual, an *unus* ("one"), impossible, but it does

---

7. *hic generatimst* Munro: *his generatim* QV: *his generat in* O: *est generatim* Lachmann: alii alia.

8. The concluding *exempla* of the second book (e.g., the cultivator of vines) reinforce this generic quality, presenting figures only identified vaguely by their activity.

seem to append a *quisque* ("each") to any usage of *unus*, bleaching distinctions between differing entities. As Friedrich Klingner puts it, treating both this and the preceding passage together, "our world enjoys no exceptional status: the same obedience to natural law governs it as everything else . . ., it obeys the law of plurality like all things in our realm of experience . . ."[9]

The strong language here marshaled against any suggestion of human (material) uniqueness is tempered somewhat by the position Lucretius articulates several hundred lines later, near the conclusion of his third book. On the whole, the third book, particularly its final *consolatio*, agrees with the materialist sentiment of the second. One could select many statements, not unlike the following lines, that seem to continue Lucretius' unstinting assault on human uniqueness: "there is need for matter so that later ages may grow, but all these ages will follow you in perishing: no differently, then, than you, did earlier generations fall; no differently will they fall." (*DRN* 3.967–69: *materies opus est, ut crescant postera saecla,/ quae tamen omnia te vita perfuncta sequentur:/ nec minus ergo ante haec quam tu cecidere cadentque.*)[10] Yet there is at least one passage within the concluding movements of the third book that suggests there is something unique about each human individual:

> *et, si iam* ***nostro*** *sentit de corpore postquam*
> *distractast animi natura animaeque potestas,*
> *nil tamen est* ***ad nos*** *qui comptu coniugioque*
> *corporis atque animae consistimus uniter apti.*
> *nec, si materiam* ***nostram*** *collegerit aetas*
> *post obitum rursumque redegerit ut sit nunc est*
> *atque iterum* ***nobis*** *fuerint data lumina vitae,*
> *pertineat quicquam tamen* ***ad nos*** *id quoque factum,*
> *interrupta semel cum sit* ***repetentia nostri.***
> *et nunc nil* ***ad nos de nobis*** *attinet, ante*
> *qui fuimus, neque iam de illis* ***nos*** *adficit angor.*
> (*DRN* 3.843–51)

9. Klingner 1952: 4–5: ". . . unsere Welt steht unter keiner Ausnahmebedingung; dieselbe Naturgesetzlichkeit beherrscht sie wie alles andere . . ., sie steht unter dem Gesetz der Pluralität wie alle Dinge unseres Erfahrungsbereiches . . ."

10. This final statement of Lucretius' comment is difficult Latin. Although the sense seems clear enough, the syntax is ambiguous. Commentators are divided on whether *haec* (*sc. saecula*) represents the subject of a clause or an object of *ante*. See the lucid remarks of Kenney (ed.) 1984: 220 (*ad loc.*), who takes *haec* as a prepositional object (as I have done in my translation here).

*And even if the mind and soul have sensation*
*after they have been drawn away from* ***our*** *bodies,*
*it is nevertheless no matter* ***to us*** *who exist*
*uniquely fitted with the union of body and spirit.*
*Nor, though time collected* ***our*** *physical matter after death*
*and configured it as it lies now, and again*
*the lights of life were given* ***to us,***
*this deed, too, would nonetheless not pertain* ***to us,***
*as soon as* ***our ability to recollect*** *(repetentia)* ***ourselves*** *has been interrupted.*
*Now nothing reaches* ***us about ourselves,*** *who we were before,*
*and no distress now impacts* ***us*** *from those (distant) things.*

At first glance, these lines seem hardly different from others that stress our own finite, material nature. The endless cycle of birth and death depicted at the conclusion to the second book, it seems, re-enters discussion here. But in once again articulating this ever-same cycle, there appears something of a struggle in identifying who "we" (*nos*) are.[11] In creating a distance between "us" and "our" *post mortem* bodies, Lucretius nonetheless still calls the bodies "ours" (843), speaking also of "our physical matter" (*materia nostra*, 847), as if—contrary to the very position he is articulating—"we" experience some attachment to the matter that once comprised "us."[12] Such a reading leans heavily upon possessive adjectives, but it dramatizes—at the lexical level—the difficulty, repeatedly met throughout Lucretius' poem, of drawing firm boundaries, whether physical or otherwise, for the individual.[13]

One central tension within the passage is enacted through the collision of a subjective with an objective perspective.[14] "We," as agents, look upon "us,"

---

11. Although she directly discusses neither this passage nor the issue of individual boundaries, Monica Gale (2004: 53) recognizes (working from a narratological perspective) a similar tension between individual and generic: "… it is a central element in the *consolatio* at the end of book 3 that the death of the individual is simply part of an endless process of atomic recycling. Millions of others have died before you, millions of other deaths will follow yours.… On the other hand, it is rhetorically important that the reader should feel directly and personally involved in these impersonal processes."

12. Compare the similar passage at 3.870–93, especially 886–87, where the personal pronouns enact the difficulty of separating one's former body from one's present self.

13. Two scholarly works relevant to Lucretius' working out of questions of limits and boundaries are De Lacy 1969 and Segal 1990: 94–114.

14. See Long 2006: 202–20, especially 218–20, where he discusses Thomas Nagel's "objective self." (This piece was published originally as Long 1997.)

as objects. This tension is felt particularly within a key phrase in the passage, *repetentia nostri* (851), translated above as "our ability to recollect ourselves."[15] *Repetentia* here names the subject's activity (or ability), while *nostri*, an objective genitive, designates the object. It is a difficult phrase for many reasons, not least because it places bounds upon an individual (or subject) through attention to that individual's own imagined limitations. That is, as traditionally understood and as here translated, *repetentia nostri* refers to a capacity for memory, a capacity to recall one's self (as a self) that is disrupted by the physical dispersion of the atoms that comprise our bodies. There are enough layers of "seeing" here to rival a hall of mirrors: "we," still alive, in confronting our own future death, assume the vantage point of that death only to discover that "we," revivified at some still more distant future moment, could not be the same ("our" selves) because our capacity for recollection would have been disrupted.

Several intriguing presumptions are at work in the passage. For one, there is the continuing presumption that "we" are all the same before the "boundary stone" of life.[16] More interesting, though, is the implicit belief that we are something more than our material selves. To be clear: *repetentia nostri*, it seems, names something other than a purely material capacity, or else it would be possible for "us" to exist, with all of our materially inscribed memories, at a future time. In other words, memory, if that is what *repetentia* is, is more than a mere physical register of our past experiences. The explanatory lines that follow the preceding passage offer some development of this position:

*nam cum respicias immensi temporis omne*
*praeteritum spatium, tum motus materiai*
*multimodis quam sint, facile hoc adcredere possis,*
*semina saepe in eodem, ut nunc sunt, ordine posta*
*haec eadem, quibus e nunc nos sumus, ante fuisse.*[17]

---

15. Although the word *repetentia* occurs only here in classical Latin, twentieth-century editors have both accepted its correctness (as a manuscript reading) and left its sense unquestioned. Ernout-Robin (1926: 2.131) proclaims: "la forme et le sens sont parfaitement clairs"; Bailey (1947: 1137) translates "the recollection of ourselves," dismissing arguments for emendation (to *retinentia*) by Lachmann; Kenney (1984: 196) renders "the ability to recollect ourselves" (as I have done earlier). Both Kenney and Bailey cite the occurrence of the term in the imitator Arnobius as positive evidence of its appearance here.

16. This "we" is all the more fascinating because it seems to join "us," as readers, to the poet, the same man who helps us to transcend "our" present perspective by imagining our own death and re-birth.

17. This line (858), first placed here by Lachmann, appears as line 865 in the manuscripts. All recent editions reproduce this transposition.

*nec memori tamen id quimus reprehendere mente:*
*inter enim iectast vitai pausa, vageque*
*deerrarunt passim motus ab sensibus omnes. (DRN 3.854–61)*

*When you look back upon the whole space of measureless time*
*that has passed, and how various the movements of matter are,*
*you can easily believe this: these very atoms (semina), from which*
*we now exist,*
*were often before placed in the same order as now.*
*Nevertheless, we are unable to grasp this fact with a recollecting mind;*
*for a break entered into our lives, and all movements*
*have wandered here and there, far afield from our senses.*

Here the words *memori ... mente* ("a recollecting mind") and *inter ... iectast vitai pausa* ("a break entered into our lives") seemingly replay *interrupta ... repetentia nostri* ("our ability to recollect ourselves ... interrupted"). The assertion seems to be that when a break (*vitai pausa*) enters into our lives (i.e., when we are destroyed), our memories are wiped clean, thus putting an end to "our" selves. That is, "our" selves are not, or not only, dependent upon our material composition (including the material composition of our minds and memories), but also upon the continual present-time functioning of our memories. Being an individual, a "self," is not merely being a material thing but *acting continually* in the full retrospective knowledge of one's self as a self.

Such a description bears some resemblance to certain modern, or post-modern, conceptions of the self. Jacques Derrida, for example, who served as one of our guides in the first chapter of this study, has articulated what may be called "a performative self," a self that is not pre-existing but rather constituted by (that is, constituted in the *act* of) utterances such as "I do" and "I promise":

> The [*sc.* Derridean] performative ... far from depending ... on a preexisting ego, I, or self, ... creates the self ..., in a way that is anticipated by Austin's extraordinary statement: "As official acts, a judge's ruling makes law; a jury's finding makes a convicted felon." (Austin 1975: 154)[18]

18. Miller 2007: 231. This article aims to disambiguate various remarks made about performatives by John Austin, Jacques Derrida, and Judith Butler.

The Derridean self described here by J. Hillis Miller, a self that is constituted and defined in and through language even if it is not a purely linguistic self, may seem initially at odds with the Lucretian self, as just described. Derrida's self appears rather immaterial or a-material, while Lucretius' is, although—as just argued—not fully material, first and foremost a consistent, material (that is, atomic) compound.[19] Yet there are stronger correspondences than these initial appearances suggest.

One of the most obviously troubling features for the Lucretian self is how to articulate it in language. Lucretius, as we have seen, repeatedly calls "ours" precisely that which he is at pains to say is not "ours." While such an observation hardly means that Lucretius defines the self linguistically, or even that he stakes an explicit position on the linguistic construction of the self, it certainly indicates that the role of language, and semiotic acts more generally, in delimiting a self is a problem for his text no less than for Derrida. It seems, moreover, that the Lucretian self is constituted contextually, which is not to say that it is arbitrary but rather to suggest that—insofar as it maintains a kernel that is not materially determined—it bears the stamp, continually and without material distinction, of its present.[20] While the structures of Lucretius' natural world, including those that govern the human individual, often appear strikingly rigid, they can, when pressed, reveal religious and other ritual (that is, performative) acts. In other words, despite the appearance of rigidity, the structures of Lucretius' natural world, as we have seen already with *foedus* and as we shall see again, are figured as "naturalized" human practices that, even if not driven by capricious acts of will, are subject to human interpretations and ideologies. The next section investigates a certain "natural" Lucretian metaphor, the *terminus* ("boundary stone"), with the aim of showing the performative (and political) acts that buttress it.

Before moving on to study the image of the *terminus*, however, it is perhaps worth re-iterating the claims advanced and explored in this section. In a selection of passages concentrated in the second and third books of the *De*

19. This description in fact verges on mischaracterizing Derrida's position, as speech acts always act within some material context, but our casual acquaintance with language tends to under-emphasize its materiality.

20. It is worth noting here Lucretius' remarks on time: "Time ... does not exist per se, but sensation arises from the things themselves ..." (*DRN* 1.459–60: *tempus* ... per se *non est, sed rebus ab ipsis/ consequitur sensus* ...) This comment is relevant precisely because the distinction between our current selves and a (possible) future material duplicate lies precisely in a temporal distinction. That is, in the latter case, the temporal (and material) context has been rendered different, which renders us (and our memories and experiences) different.

*rerum natura*, we have traced how Lucretius, while emphasizing the generic, material nature of human individuals, nonetheless appears to reserve space for the uniqueness of their particular experiences and memories. This location of the human individual at the intersection of a general, all-encompassing materialism and a more subjective set of personal experiences defines the *unus quisque*, who will be an important figure in our study of Lucretius' anthropology and, more specifically, the origins of language within that anthropology.

## (ii) Boundaries (for Many)

A key passage in constructing the foregoing argument, *DRN* 2.1077–89, stresses the broad application and inevitability of natural processes for different *genera*. Humans, like fish, like birds, will perish. Klingner articulates this inevitability in the terms of "natural law" (*Naturgesetzlichkeit*), a metaphor hardly unfamiliar to Lucretius himself (e.g., *foedus naturae, lex naturae*), but in this particular passage the poet deploys a more physical metaphor, that of the *terminus* ("boundary stone, a marker of individual property lines"), to denote the rigid fixity of death across species.[21] The word *terminus* occurs at various turns throughout the poem, enough to be a prominent, if not dominating, metaphor.[22] It occurs not only in stressing firm division but also in defining power and its distribution. We may recall, for example, the lines early in Lucretius' work that describe what the victorious Epicurus offers us from his mental conquest of traditional religious superstition:

> *... refert nobis victor quid* ***possit*** *oriri,*
> *quid* ***nequeat,*** *finita* ***potestas*** *denique* ***cuique***
> *quanam sit ratione atque alte* ***terminus haerens.***
> *quare religio pedibus subiecta vicissim*
> *obteritur, nos exaequat victoria caelo. (DRN 1.75–79)*

> *... he victoriously brings back to us what* ***can*** *arise,*
> *what* ***cannot,*** *in which way, finally,* ***power*** *is defined*

21. For a fuller discussion of the term *foedus*, specifically within the phrase *foedus naturae*, and the role it plays within Lucretius' text, see Garani 2007, Asmis 2008, and Chapter 2 of this study.

22. To be precise, *terminus* occurs six times in the poem, at *DRN* 1.77, 1.596, 2.1087, 3.1020, 5.90, 6.66, all of which are considered in the current discussion. It is perhaps worth noting here that Lucretius was certainly not the first to employ the metaphor, as a fragment of Accius indicates: "the old boundary stone of the fates had commanded thus" (*trag.* 481: *veter fatorum terminus sic iusserat*).

> ***for each*** *and its* ***deep-set boundary stone.***
> *Thus, religious superstition, cast under our feet,*
> *is trampled, victory brings us up to the heavens.*

This deployment of *terminus*, while apparently carrying the same stress on fixity and limit as *terminus vitae*, is more general in scope, seemingly addressing itself to the whole range of limits within the nature of things. The connection fashioned here between potential (*possit, nequeat, potestas*), or power, and *terminus* also begins, if only in a oblique fashion, to articulate a political claim—how power (*potestas*) is defined for each (*quisque*)[23]—that will be explored further as the present section (and the remainder of the chapter) develops. As a first step in that process, it will be useful to probe the term *terminus* further.

However much it may offend modern, skeptical sensibilities, *terminus*, the "boundary stone," was regularly understood in our ancient sources not as an inanimate stone but rather as one inhabited by a *numen* ("godhead; *lit.* a [divine] nod"), *Terminus*.[24] This divinity was celebrated, among other means, by the festival of the Terminalia on 23 February.[25] Indeed, some sources, like the lexicon of Festus (as it is transmitted to us through Paul the Deacon), focus more or less exclusively on the *numen* at the expense of any other possible significations of *Terminus/terminus*. Festus preserves two entries for *Terminus*, both of which aim to explicate aspects of the *numen*:

> **Termino** sacra faciebant, quod in eius tutela fines agrorum esse putabant. denique Numa Pompilius statuit, eum, qui terminum exarasset, et ipsum et boves sacros esse. (Festus 505.19–21 [Lindsay])

> **In honor of Terminus**, they used to perform rites, because they thought the boundaries of the fields were under his protection. Moreover,

23. Bailey *ad loc.* (1947: 612) rightly notes "the legal ring" of *finita potestas*, as if Lucretius were describing equality before the law rather than physical properties.

24. Wissowa 1912: 136–38, provides a brief account references to the primary source material on the divinity Terminus. A more recent treatment, with extensive consideration of both Roman and comparative evidence, is Piccaluga 1974. On *numina* more generally, see BNP 1.10–11; 2.3–4 and the bibliography presented there.

25. The date of the Terminalia is clear from Ovid *Fasti* 2.639–84 (February 23), a passage discussed elsewhere in this chapter.

> Numa Pompilius established that, for him who had plowed up the boundary stone, both the man himself and his cattle were sacred.[26]

> **Terminus** quo loco colebatur, super eum foramen patebat in tecto, quod nefas esse putarent, terminum intra tectum consistere. (Festus 505.22–24 [Lindsay])

> **Terminus**: above the place where he was worshipped, a hole in the roof lay open, because they thought it sacrilege to establish a boundary stone beneath a roof.

It is no surprise that a divinity of the boundary stone is memorialized primarily as a god who divides space, one who marks boundaries and demands to be free of roofs. But the acts that accompany this division of space, whether aimed at its preservation or its destruction, are, or at least may be read as, symptoms of larger political claims. The impulse to preserve boundaries suggests not only an obsession with order but also a certain fairness toward one's neighbors. The impulse to trample boundaries, by contrast, suggests a nascent imperialism, or, at a minimum, several ancient sources read the impulse in this fashion.

Plutarch, for one, dramatizes these impulses—preservation of boundaries against their trampling (or even their non-existence)—through the two mythic figures of Romulus and Numa Pompilius. In his *Roman Questions*, the Greek moralist and philosopher explores the question of why Terminus (reportedly) was originally honored without blood sacrifice:

> "Διὰ τί τὸν Τέρμινον, ᾧ τὰ Τερμινάλια ποιοῦσι, θεὸν νομίζοντες οὐδὲν ἔθυον αὐτῷ ζῷον;" ἢ Ῥωμύλος μὲν ὅρους οὐκ ἔθηκε τῆς χώρας, ὅπως ἐξῇ προϊέναι καὶ ἀποτέμνεσθαι καὶ νομίζειν πᾶσαν ἰδίαν, ὥσπερ ὁ Λάκων εἶπεν, ἧς ἂν τὸ δόρυ ἐφικνῆται· Νομᾶς δὲ Πομπίλιος, ἀνὴρ δίκαιος καὶ πολιτικὸς ὢν καὶ φιλόσοφος γενόμενος, τήν τε χώραν ὡρίσατο πρὸς τοὺς γειτνιῶντας καὶ τοῖς ὅροις ἐπιφημίσας τὸν Τέρμινον ὡς ἐπίσκοπον καὶ φύλακα φιλίας καὶ

26. On this practice, reportedly established by Numa, see Dion. Hal. *AR* 2.74.3. Saying that a man who disturbed a boundary stone was "sacred" (*sacer*) apparently means, among other things, that he can be killed with impunity. "The *homo sacer* is he, upon whom the people have passed judgment because of a misdeed; it is not right to kill him, but whoever kills him is not condemned as a murderer." (Festus 424.5–8: *At homo sacer is est, quem populus iudicavit ob maleficium; neque fas est eum immolari, sed qui occidit, parricidi non damnatur* ...) Cf., too, Giorgio Agamben's treatment of the *homo sacer* at Agamben 1998: 71–115, which sets up his treatment of modern biopolitics.

> εἰρήνης ᾤετο δεῖν αἵματος καὶ φόνου καθαρὸν καὶ ἀμίαντον διαφυλάττειν; (Plutarch *QR* 15 [*Moralia* 267C])
>
> "For what reason did they (*sc.* the Romans) formerly sacrifice no living creature to Terminus, in whose honor they hold the Terminalia, although they consider him a god?" Is it because Romulus did not place limits upon his land (in order that, as the Spartan said,[27] it might be possible to go forward and take [*lit.* cut off for themselves] and believe the land entirely theirs, wherever their spear reaches), whereas Numa Pompilius, a just man who was both a statesman and a philosopher, divided the land towards their neighbors and proclaimed Terminus the overseer of the boundaries and the guardian of friendship and peace, thinking it proper to preserve him pure and unstained by blood and murder?

Here Romulus, even if he does not transgress boundaries, avoids fashioning them out of his reluctance to restrain the military might of the Romans. Numa, by contrast, "a just man" (ἀνὴρ δίκαιος), establishes boundaries and sets up Terminus to watch over them, aiming to preserve friendship and peace. This contrast, which draws upon stock characterizations of these two mythic Roman figures, provides a lesson in the ways in which *Terminus/terminus* was conceived and figured politically.[28] The divinity is located here, as elsewhere, on the side of justice, friendship, and peace, while at the same time standing at the threshold of a violence and imperialism represented by Romulus. Romulus, like Terminus, was firmly associated with boundaries: he is recorded as author of the Roman colonization ritual, the *deductio* (the "leading out" of the colony), which marked out the boundaries of the new city with plowed furrows.[29] If Terminus stands as a figure for the preservation of boundaries, Romulus stands rather as a figure for the imperial (and imperialist) task of

27. On the comment of "the Spartan," compare Plutarch *Mor.* 210E: Ερωτηθεὶς δέ ποτε ἄχρι τίνος εἰσὶν οἱ τῆς Λακωνικῆς ὅροι, κραδάνας τὸ δόρυ εἶπεν "ἄχρις οὗ τοῦτο φθάνει." ("Once he was asked 'How far do the boundaries of Sparta extend?' He shook his spear and said 'As far as this outstrips the competition.' ")

28. See, e.g., T. P. Wiseman's recent characterization (in Wiseman 2004) of Numa Pompilius as "[t]he Romans' paradigm of a wise and pious king" (51). He speaks, by contrast, of Romulus' "confidence and speed of action" as well as his reported establishment of "the ritual procedure used ... for ... colonial foundations" (141).

29. On the *deductio*, see Varro *LL* 5.143, Ovid *Fasti* 4.825 (for Romulus performing the original *deductio* at Rome). Modern accounts include (an extremely abbreviated list) Kornemann 1901: 511–88, MacKendrick 1952: 140–41, and Salmon 1969: 20–25.

creating and re-drawing the boundaries of the ever-expanding territories of Rome.

Plutarch's characterization of Terminus is continued in many ways by the representation of that same *numen* in Ovid's *Fasti*. On the date of the Terminalia (February 23), the festival held in honor of Terminus, Ovid depicts a rustic celebration in which the divinity is called "communal Terminus" (2.655: *communis Terminus*)[30] and championed by the "simple neighborhood" (657: *vicinia simplex*) as if he were a morally upstanding politician:

> *tu populos urbesque et regna ingentia finis:*
> *omnis erit sine te litigiosus ager.*
> *nulla tibi ambitio est, nullo corrumperis auro,*
> *legitima servas credita rura fide. (Ovid Fasti 2.659–62)*

> *You mark out peoples and great kingdoms:*
> *every field would be contested without you.*
> *You cannot be bribed, no gold corrupts you,*
> *you preserve entrusted lands with genuine faith.*

This praise of Terminus borders on the absurd (especially as the *numen* was never depicted anthropomorphically[31]), but it nonetheless eloquently capitalizes upon a genuine perception of the divinity as communal and just. Indeed, as Ovid's hypothetical speech goes on, Terminus is characterized as a unique holdout to the imposition of Jupiter's new rule (that is, his new temple) on the Capitoline hill:

> *quid, nova cum fierent Capitolia? nempe deorum*
> *cuncta Iovi cessit turba locumque dedit;*

---

30. Compare the assertion of the historian Dionysius of Halicarnassus that Terminus divides public from private:

> τοῦτο δ᾽ οὐκ ἐπὶ τῶν ἰδιωτικῶν κατεστήσατο μόνον κτήσεων τὸ δίκαιον, ἀλλὰ καὶ ἐπὶ τῶν δημοσίων, ὅροις κἀκείνας περιλαβών, ἵνα καὶ τὴν ῾Ρωμαίων γῆν ἀπὸ τῆς ἀστυγείτονος ὅριοι διαιρῶσι θεοὶ καὶ τὴν κοινὴν ἀπὸ τῆς ἰδίας. (Dion. Hal. 2.74.4)

> Further, he (*sc.* Numa) established this justice (*sc.* that a mover of boundary-stones was a *homo sacer*) not only for private properties but also for public ones. He encircled those properties with boundary markers, too, in order that the gods of the boundaries might divide both the land of the Romans from their neighbors' and public land from private land.

31. See BNP 2.3–4 (1.1b) and the references given there.

*Terminus, ut veteres memorant, inventus in aede*
*restitit et magno cum Iove templa tenet.*
*nunc quoque, se supra ne quid nisi sidera cernat,*
*exiguum templi tecta foramen habent.*
*Termine, post illud levitas tibi libera non est:*
*qua positus fueris in statione, mane;*
*nec tu vicino quicquam concede roganti,*
*ne videare hominem praeposuisse Iovi:*
*et seu vomeribus seu tu pulsabere rastris,*
*clamato, "tuus est hic ager, ille tuus." (2.667–78)*

*What happened when the Capitoline was transformed? Surely the whole crowd*
*of gods yielded to Jupiter and granted him place.*
*As our elders report, Terminus, discovered in the shrine,*
*resisted, and he holds the sanctuary with Jove.*
*Now, too, lest he discern anything except stars above him,*
*the roof of the shrine has a small opening.*
*Terminus, after that, fickleness is not free for you:*
*abide in the post where you've been placed;*
*don't yield a thing to your demanding neighbor,*
*lest you seem to rank a human before Jupiter:*
*and whether you will be stricken with ploughshares or drag-hoes,*
*exclaim, "This is your field; that, yours."*

These lines further strengthen, if possible, the depiction of Terminus as both just and anti-imperialist. Although other versions have it differently, within the context of the present poem he is the lone divinity to resist Jupiter's tyranny;[32] and that resistance now limits his freedom. He must not, according to Ovid's advice, yield ground to any human for fear of seeming to rank him above Jupiter. In other words, Terminus enacts not only a division of space but also (implicitly)

32. The notes (*ad loc.*, 201–2) to the Penguin translation of Ovid's *Fasti* (A. J. Boyle and R. D. Woodard, eds.) report: "Roman tradition held that Terminus, along with Juventas, the goddess of youth . . ., and many other deities had shrines on the Capitoline Hill which had to be removed to make room for the great temple of Jupiter Optimus Maximus. While most of the gods gave way, Terminus and Juventas refused to vacate; in the end, shrines for both were incorporated into the new temple." These remarks essentially distill the narrative at Dion. Hal. *AR* 3.69.5. Thus, although he is alone in Ovid's account, in other accounts Juventas accompanied him in his resistance to Jupiter. Both divinities became cultic aspects of Jupiter. See Wissowa 1912: 135–36 on Juventas.

a separation of classes or species, of human and divine. Moreover, the narrative depicts Terminus' primary activity, the preservation of boundaries, precisely as a performative act (declaring "*This* is your field; *that*, yours" ["tuus est hic ager, ille tuus"]) rather than a default condition. The fact that Ovid renders his advice with a future imperative (*clamato*) seems to represent Terminus not as a passive marker of boundaries but as one who constantly performs and will continue to perform a dividing function. Such observations are suggestive for our reading of Lucretian imagery and the significance of his "deep-set boundary stone" (*terminus haerens*), but before returning to Lucretius, it is worth reading to its conclusion Ovid's commentary on the Terminalia:

> *est via quae populum Laurentes ducit in agros,*
> *quondam Dardanio regna petita duci:*
> *illa lanigeri pecoris tibi, Termine, fibris*
> *sacra videt fieri sextus ab Urbe lapis.*
> *gentibus est aliis tellus data limite certo:*
> *Romanae spatium est Urbis et orbis idem.*[33] *(2.683–84)*
>
> *There is a path that leads people into the Laurentine fields,*
> *kingdoms once sought by the Dardanian leader (i.e. Aeneas):*
> *its sixth stone from Rome sees rites occur for you,*
> *Terminus, with the entrails of a wool-bearing sheep.*
> *For other peoples, land is given with a fixed boundary:*
> *the domain of the Rome is the world.*

Even in the absence of any explicit mention of Romulus, it seems that a certain imperialism, a certain ability to transcend boundaries, accompanies any appearance of Terminus.

Turning, then, to the text of Lucretius, we may quickly review a few instances of the term *terminus*. While there is no explicit reference to the divinity Terminus anywhere within *De rerum natura*, many features associated with the *numen* lurk nonetheless:

> ...
> *inmutabili' materiae quoque corpus habere*
> *debent nimirum; nam si primordia rerum*

33. See Oliensis 1998: 107–27 (esp. 107–11 of the section "Lyric and empire") who reads these lines in treating *Terminus/terminus* and the political significance of boundaries within several Horatian odes.

*commutari aliqua possent ratione revicta,*
*incertum quoque iam constet quid possit oriri,*
*quid nequeat, finita potestas denique cuique*
*quanam sit ratione atque alte* ***terminus haerens,***
*nec totiens possent* ***generatim*** *saecla referre*
*naturam mores victum motusque parentum. (DRN 1.591–98)*

. . .
*moreover, they*[34] *ought to have, no doubt, a body*
*of immutable matter; for if the first-beginnings of things*
*could be changed about, overcome in some fashion,*
*it would then be uncertain what could arise,*
*what could not, how power and the* ***deep-set boundary stone***
*are defined for each, and the races,* ***generation by generation,***
*would not be able to reproduce so often the nature—*
*the habits, the lifestyle, and the movements—of their parents.*

This passage, which repeats the language of Epicurus' discoveries (1.75–77), places *terminus* at the intersection of features we traced in the preceding section.[35] The boundary stone, it is stressed, is a generic marker that acts on the level of groups (*generatim saecla*), but it also—as the re-iterated lines suggest—defines the individual (*finita potestas . . . cuique*). Its broad yet particular application conjures up some of the (previously examined) democratic character of Terminus, a just guardian of boundaries for all, although the chief effect of this passage within its context is to stress a continual sameness. Children are like their parents, a sign that we are all comprised of the same immutable matter (*inmutabilis materiae*).

These lines seem to linger on the level of the physical world, without teasing out, at least not in any extended fashion, the potential political (or ethical) implications of such physical properties. Yet there are other passages where Lucretius makes more explicit connections between physics and politics. In other words, as with the Plutarch and Ovid passages already considered, divisions of material space take on significance beyond the merely physical. One of the culminating moments in Lucretius' tirade

34. The immediate grammatical subject here is *variae volucres* ("different kinds of birds"), but the tone of the passage (including its location within an argument about atoms generally) suggests that the comment is intended to apply quite broadly.

35. The language of the passage, through its verbal repetition of earlier phrasing (that is, through its lexical sameness), perhaps carries part of the burden of its argument, which addresses the material sameness and continuity of the world.

against the fear of death, for example, lends death an essentially democratic character:

*sed metus in vita poenarum pro male factis*
*est insignibus insignis, scelerisque luella—*
*carcer et horribilis de saxo iactu' deorsum,*
*verbera carnifices robur pix lammina taedae;*
*quae tamen etsi absunt, at mens sibi conscia factis*
*praemetuens adhibet stimulos torretque flagellis,*
*nec videt interea* ***qui terminus*** *esse malorum*
*possit nec quae sit poenarum denique finis,*
*atque eadem metuit magis haec ne in morte gravescant.*
*hic Acherusia fit stultorum denique vita.*
*hoc etiam tibi tute interdum dicere possis:*
***'lumina sis oculis etiam bonus Ancu' reliquit,***[36]
***qui melior multis quam tu fuit, improbe, rebus.***
***inde alii multi reges rerumque potentes***
***occiderunt, magnis qui gentibus imperitarunt.*** *' (DRN 3.1016–28)*

*But fear of punishment for notorious misdeeds in life*
*is notorious, and paying for the crime (sc. is notorious, too)—*
*prison and being thrown down from a rock frightfully,*
*whips, executioners, the rack, pitch, the metal plate, torches.*
*Even when these things are absent, the mind, conscious of its own deeds,*
*in its fear applies prods and burns with lashes,*
*and, moreover, it sees neither* ***what*** *could be the* ***limit*** *to ills,*
*nor what end there could be to punishments,*
*and it feared these same things all the more, lest they grow more severe in death.*
*Here, then, death in life occurs for the ignorant.*
*This, too, you could sometime say to yourself:*
***"Even good Ancus***[37] ***let the light out from his eyes,***
***who was much better than you, shameless man, in many things.***
***Many other kings, ruling over things, have fallen since then,***
***kings who used to command great peoples."***

---

36. This line seems to be borrowed, more or less directly, from Ennius' *Annales* (= 137 Skutsch). For further discussion, see Skutsch 1985: 293–95.

37. Ancus here refers to Ancus Marcius, the fourth king of Rome, who was "always the archetypal 'good king'" (Kenney 1984: 233).

Here death serves as a leveler. Not only, as we have seen previously in our study of Lucretius' second book, does death impact all *genera* in the same fashion, but it impacts all individuals within a given *genus* alike. After railing against the fear of punishment (in life and death) for several lines—lines that spend more time detailing horrid punishments than offering any perceptible solace—Lucretius turns to cataloguing great individuals who have perished. The list, which is only cited in its first several lines here, ranges from the "good" King Ancus to Scipio Africanus[38] to Epicurus and Democritus. And while the larger thrust of the passage no doubt retains a focus, in the traditionally harsh language of diatribe, on unmasking misconceptions about death and the anxieties surrounding it, the portrayal of death as an egalitarian agent, who receives all comers, both kings and paupers, equally, well evokes some of the earlier, democratic depictions of Terminus.

A passage in the sixth book develops the boundary stone rather as a defense against the perils of monarchy, at least in the figurative terms of the passage:

*cetera quae fieri in terris caeloque tuentur*
*mortales, pavidis cum pendent mentibu' saepe,*
*et faciunt animos humilis formidine divom*
*depressosque premunt ad terram propterea quod*
***ignorantia causarum conferre deorum***
***cogit ad imperium res et concedere regnum.***
. . .
*nam bene qui didicere deos securum agere aevom,*
*si tamen interea mirantur qua ratione*
*quaeque geri possint, praesertim rebus in illis*
*quae supera caput aetheriis cernuntur in oris,*
*rursus in antiquas referuntur religiones,*
*et **dominos acris** adsciscunt, omnia posse*
*quos miseri credunt, **ignari** quid queat esse,*
*quid nequeat, finita potestas denique cuique*
*quanam sit ratione atque alte **terminus haerens;***
*quo magis errantes caeca ratione feruntur. (DRN 6.50–55, 58–67)*

*All things that mortals observe happening on earth*
*and in the sky, when often they rely upon their fearful minds,*

38. Or more precisely, *Scipiadas* (3.1034), "the scion of the Scipios," who seems to be the elder Scipio Africanus.

*these, too, make their spirits low out of fear of the gods*
*and they push them, already down, to the earth because*
***their ignorance of causes drives them to grant things***
***to the supreme command of the gods and yield the kingdom.***
. . .
*For, those who have learned well that gods lead a secure life,*
*if nevertheless in the meantime they wonder in what fashion*
*all things can be conducted, especially in those things*
*which are discerned in the skies above our heads,*
*they fall back into old superstitions,*
*and they adopt **harsh leaders**, whom they, wretched,*
*believe to be all-powerful, **unaware** what can be,*
*what cannot, how power and **the deep-sticking***
***boundary stone** are defined for each;*
*thus they wander all the more, carried on by blind reason.*

These lines, situated near the beginning of the sixth book, offer, in a sense, a recapitulation of the narrative of the rise of religion (*religio*), as told in greater detail within Lucretius' fifth book. The present examination will turn shortly to study that treatment, but it is worth pausing for a moment to observe how directly this passage states the political implications of the seemingly physical knowledge offered by Lucretian (and Epicurean) wisdom. Attributing natural causation to the gods (through ignorance of the true, Epicurean account) is yielding to their supreme command (*imperium*) and adopting harsh leaders (*dominos acris*). In other words, in the implicit and explicit terms of these figurations, there is a political component, a resistance to a certain tyranny (or oligarchy) of the gods, in using our own senses to study the heavens properly. These politics are only suggested in their barest outlines here, but even this bare suggestion, when coupled with the other evidence we have seen, is enough to establish that *terminus* conjures up a certain, albeit complex, democratic quality.

The very lines that conclude this passage (6.58–66) are, word for word, one of Lucretius' opening salvos (5.82–90) for announcing his program in the fifth book. Earlier examination within this study has suggested that the fifth book, at least in certain moments, is a book that emphasizes creation and depicts the poet himself as, in the words of Godo Lieberg, *poeta creator*. Creation will therefore take center stage in the next section, a section that focuses upon elaborating connections between the politics we have begun to articulate in the first two sections of this chapter and language, as described in Lucretius' anthropology. More specifically, in bringing back into focus acts of linguistic creation—that is, speech acts—the next section elaborates a "catachrestic potential" in language but also in Lucretius' anthropology more generally.

## II. Catachreses: Language and Divine Absence

This section draws upon the analyses of preceding sections, taken together with a reading of passages culled from Lucretius' fifth book, to suggest that Lucretius formulates, in the first instance, a "catachrestic potential" that at once defines the power of language and describes the political activities (i.e., the overthrow of kings, the rise of *religio*) narrated within his anthropology. In other words, continuing the line of analysis begun previously, we here attempt to demonstrate that certain (apparently) physical principles (i.e., equality before/fixity of laws of nature, atomic motion) move between physical and political spheres. To be more explicit, this section argues for a connection between the (already partially explored) emphasis on *unus quisque* and an anti-monarchical, "catachrestic" politics. Each time—whether in cosmology, politics, or language—forces conspire to create a single, monarchical governance, Lucretius' text moves to undermine and question the authority of that monarchy, instead replacing that monarchy with a more chaotic, democratic political structure.

### (i) Natural Catachreses

Perhaps the best place to begin this discussion is with nature (*natura*) herself. Numerous scholars have commented upon the fashion in which Lucretius uses *natura*, which is so often personified (*Natura*) within his text.[39] It has been a standard criticism of Epicurean thought since antiquity that it excludes the traditional gods from explanation of the causal functioning of the heavens only to replace them with a new god called "Nature."[40] Even sympathetic critics have struggled

39. Instances of personification are too numerous to list here with any ease, but—within the fifth book alone—examples occur at *DRN* 5.77, 186, 206–7, 225, 234, 831, 846, 871, 1028. See the fuller catalogue, with classification, at Fowler 2002: 243. For further discussion of the various meanings of *natura* within Lucretius, see Campbell 2003: 80–82 (with the remarks of Holmes 2005b), Sallmann 1962, and the appendix to Lovejoy and Boas, "Some Meanings of 'Nature'" (Lovejoy and Boas (eds.) 1935: 447–56), where sixty-six (66) distinct meanings are catalogued. Also of related interest are Clay 1969 (with his emphasis on the creative power of Lucretian *natura*) and Kennedy 2000a: 218–19 (on personification).

40. See the fairly typical remarks of Minucius Felix: "Even the famed Epicurus, who imagines either that gods are unnecessary or that they do not exist, puts Nature up top." (*Octavius* 19.8–9: *Etiam Epicurus ille, qui deos aut otiosos fingit aut nullos, Naturam superponit.*) It should

to find the appropriate language to describe the function of such extensive personification. Don Fowler, for one, writes that *natura* "replaces" the gods "but as a place-marker, semantically transparent like the *it* in 'it is raining.'"[41] Klaus Sallmann, author of a lengthy article on the meaning of *natura*, comments similarly that *natura* stands as a "surrogate for the world of the gods (*Surrogat der Götterwelt*), which—although it occupies their evacuated position—nonetheless does so in a rather different and un-divine fashion (*ganz anders und ungöttlich*)."[42]

These interpretations are important for present considerations precisely because *natura* is not only personified but also often presented as a ruler or governor. Early in the fifth book, for example, Lucretius remarks that he "will lay out by what power governing nature steers" (*DRN* 5.77: *expediam qua vi flectat Natura gubernans*).[43] Elsewhere he comments, "Nature changes all things and compels them to turn" (831: *omnia commutat Natura et vertere cogit*). Such statements suggest that if one aims to investigate the organizational structure of the cosmos (which may serve in turn as a model for a political structure[44]), one would do well to grapple with Lucretius' representation of *natura* directly. If, then, in attempting to parse *natura*, we read the comments of Fowler and Sallmann closely, it is striking that they both describe *natura* in terms of location and movement. Fowler speaks of *natura* re-*placing* the gods, and he calls *natura* a *place*-marker. Sallmann, even more explictly, writes of how *natura* occupies the "evacuated *position*" (*leergewordene* Stelle) of the gods in Lucretius' poem. These analyses tend to suggest a theme that is developed in greater detail as the present discussion unfolds: *natura* is what we may call a "catachrestic" divinity.

---

be noted that personifying nature (φύσις, *natura*) is predominently, if not entirely, a trope of Lucretius rather than of Epicurus. (One exception, sometimes ascribed to Metrodorus rather than Epicurus, is Usener 469: "Thanks to blessed Nature, because she made the necessary easy, and the difficult unnecessary." [χάρις τῇ μακαρίᾳ Φύσει, ὅτι τὰ ἀναγκαῖα ἐποίησεν εὐπόριστα, τὰ δὲ δυσπόριστα οὐκ ἀναγκαῖα.])

41. Fowler 2002: 243, part of a comment on *DRN* 2.168.

42. Sallmann 1962: 281, cited at Fowler 2002: 243.

43. On the striking phrase "governing Nature" (*natura gubernans*), see Gigandet 1996.

44. Taking the (perceived) structure of the heavens as a model for human governance seems to be a common ancient trope, one that appears implictly in the rhetoric of a passage we have already seen (*DRN* 6.50–67), where the gods were compared to "harsh rulers." A more explicit statement of this move (from the cosmos to the human sphere) may be found in Cicero at *DRP* 1.56: "... so people have consented, no doubt by the decrees of their leaders, that nothing is better than the king, since they believe all the gods to be ruled by the divine will of one ..." (... *ita consensisse gentes, decretis videlicet principum, nihil esse rege melius, quoniam deos omnes censent unius regi numine* ...).

To clarify this statement, it will be useful to define catachresis more fully. Catachresis names a rhetorical trope in which, as the handbook *Ad Herennium* has it, we "abuse" a "neighboring" term for a "proper" one:[45]

> **Abusio** est, quae verbo simili et **propinquo** pro certo et proprio abutitur, hoc modo: "vires hominis breves sunt"; aut: "parva statura"; aut: "longum in homine consilium"; aut: "oratio magna"; aut: "uti pauco sermone." nam hic facile est intellectu **finitima** verba rerum dissimilium ratione abusionis esse traducta. ([Cicero] *Ad C. Herennium* 4.45)[46]
>
> **Catachresis** is that trope which "abuses" a certain similar, **neighboring** word for the fixed, proper one, as follows: "the strength of the man is 'short-lived' "; or: " 'little' stature"; or: "a man 'long on' counsel"; or: " 'great' speech"; or: "to have a 'bit of' conversation." This phenomenon is easy to grasp: **neighboring** words have been transferred to dissimilar things through catachresis.

Catachresis thus describes a *movement* from a proper sphere into an improper one. It is often identified as a subspecies of metaphor, a term that itself literally means a "transfer, or carrying over."[47] When Fowler and Sallmann speak of *Natura* re-*placing* the traditional gods, *Natura* thus becomes a catachrestic divinity insofar as it moves into a sphere, the divine, that is not properly its own.

A somewhat different definition of catachresis, found in Cicero's *De oratore* (which has already appeared in earlier discussion), brings out other connotations of the trope:

> tertius ille modus transferendi verbi late patet, quem **necessitas** genuit **inopia coacta et angustiis**, post autem delectatio iucunditasque

---

45. On catachresis generally, see Stanford 1936: 37 and Lausberg 1990: 288–91 (§ 562).

46. An intriguing comparandum for this definition is provided by Quintilian, who paints catachresis as a particularly Roman trope (8.6.34–35): "There is, therefore, all the more need for catachresis, which we correctly call *abusio*, that provides its own name that is nearest to those who do not have one." (*Eo magis necessaria catachresis, quam recte dicimus abusionem, quae non habentibus nomen suum accommodat* ***quod in proximo est***.) The context of this statement paints catachresis as the preferred alternative (over against *onomatopoeia*, "the making of names") for introducing new "terms" into Latin without actually introducing new words. Of onomatopoeia he writes (8.6.31): "for the Greeks, it was held amongst the greatest virtues; for us, it is scarcely tolerated." (*Graecis inter maximas habita virtutes, nobis vix permittitur*). It is for this reason that Romans need catachresis all the more (*eo magis necessaria catachresis*).

47. See Cicero *Orator* 94: "Under the heading of *translatio* ("metaphor; transference"), Aristotle includes … *abusio*, which he calls *catachresis* …" (*Aristoteles autem translationi … subiungit … abusionem, quam* κατάχρησιν *vocat …*)

> celebravit. nam, ut vestis frigoris depellendi causa reperta primo, post adhiberi coepta est ad ornatum etiam corporis et dignitatem, *sic* **verbi translatio** instituta est **inopiae causa**, frequentata delectationis. nam "gemmare vitis, luxuriem esse in herbis, laetas segetes" etiam rustici dicunt. quod enim declarari vix verbo proprio potest, id **translato** cum est dictum, inlustrat id quod intellegi volumus eius rei, quam **alieno** verbo posuimus, similitudo. ergo has translationes quasi mutuationes sunt, cum quod non habeas **aliunde** sumas ... (Cicero *De oratore* 3.155–56)
>
> That third manner [*sc.* of adorning an oration], that of using a word metaphorically, is common all over, which method **necessity** bore, **compelled by poverty and narrow means**, but later it was celebrated with pleasure and good will: for, as clothes were first invented for fighting off the cold, then began to be worn for ornament and status, so **metaphor** was established **due to poverty**, then was attended with pleasure; for even country bumpkins say "vines 'sparkle with buds,' " " 'riches' are in the grassy field," "the corn crops are 'happy.' " And when what can scarcely be named with a proper term is named with **a transferred one**, the similarity of that thing which we have named with a **foreign** word illuminates that which we desire to be understood. Therefore, these metaphors are like taking out loans, since you take what you don't have **from elsewhere** ...

In discussing the ornate style, L. Licinius Crassus continues to stress the movement named by the trope catachresis, its "foreignness" in drawing terms from "elsewhere." But alongside this stress on movement, there is also an emphasis on necessity (*necessitas*) and poverty (*inopia*). Crassus seems to be suggesting in this passage that catachresis, at least in its origins, was not a voluntary choice or artful "turning" of the text but rather a needed development of language, a language on its course to greater sophistication and civilization. The comparison with clothing is striking, as it conjures up a vivid image for a catachresis that begins involuntarily and yet becomes an ornamental marker of status (*dignitas*).

This narrative of catachresis as a necessity initially brought about by poverty and only later an ornamental choice allows us a glimpse of a problem that has already appeared under the guise of the figurative conflation of atoms and humans. Viewed from a human perspective (i.e., the perspective of the "atomization of humans"), a perspective that often predominates in Lucretius' fifth book, the juxtaposition of humans and atoms seems to call into question the notion of a pure human agency.[48] That is, passages scattered throughout the *De rerum natura* make the point that humans often seem to act automatically, as if following the dictates of a physical law rather than a clear volition. We may, for example, recall a passage that occurs within Lucretius' anthropology at the moment when the human race, still hard and uncivilized, begins to join together randomly in lust:

> *nec commune bonum poterant spectare, neque ullis*
> *moribus inter se scibant nec legibus uti.*
> *quod cuique obtulerat praedae fortuna, ferebat*
> ***sponte sua** sibi quisque valere et vivere doctus.*
> *et Venus in silvis iungebat corpora amantum;*
> ***conciliabat** enim vel mutua quamque cupido*
> *vel violenta viri vis atque inpensa libido*
> *vel pretium, glandes atque arbita vel pira lecta. (DRN 5.958–65)*
>
> *They were not able to consider the common good, and*
> *they did not know how to use any manners nor laws.*
> *Whatever prize fortune granted to each, he would bear*
> ***by his own power** for himself, having learned to live and fare well.*
> *And Venus used to join the bodies of lovers;*
> *for either common desire **attracted** each woman*
> *or the violent force of a man and his excessive lust*
> *or a bribe, acorns and arbute-berries or select pears.*

As we have already suggested elsewhere, it is hard to deny that these primitive scenes of mating have an atomic flavor, especially as the vocabulary (*sua sponte, conciliabat*) is the same as that used to describe atomic movements and couplings.

The developmental narrative offered by Crassus in the *De oratore* creates a similar relationship between speakers and the catachrestic language they may

48. See the remarks at Clay 1998: 169 (originally published in 1996), which in some ways anticipate the claims made here: "… [*sc.* Lucretian metaphor] can wear the face of *catachresis* [emphasis original] as it confounds and confuses the realities language has segregated into distinct categories. In contaminating the atomic with the human, Lucretius' metaphors serve an *ethical* [emphasis added] end, for they hold up the invisible conflict of blind bodies moving blindly in space and engaging in battle as an image against which the tranquility of Epicurean philosophy is the alternative." I agree with Clay's emphasis on the ethical (and further, the political) import of Lucretian catachrestic language, although I am interested more specifically in the implications of undermining intention and volition than Clay.

author. If catachrestic expressions—we may think, for example, of much of the Lucretian vocabulary of the natural world (*concilia, foedus, terminus*)—are driven by poverty (*DRN* 1.139: *egestas linguae*)[49] and necessity rather than a clear volition, then it raises not only the question of the precision of such terminology but also that of the structure (at once political and linguistic) that lurks behind its deployment.[50] That is, what classically and paradigmatically appears as one (a singular subject) disguises a whole system of compulsory material interactions—that is, in a numerical sense, many. If we return to our initial focus on *Natura*, we may note that Lucretius seems, at least at times, to suggest this very structural fact (i.e., that a diverse many may lurk behind an apparently unitary agent):

*nam tibi de summa caeli ratione deumque*
*disserere incipiam et rerum primordia pandam*
*unde omnis Natura creet res auctet alatque*
*quove eadem rursus Natura perempta resolvat;*
*quae nos "materiem" et "genitalia corpora rebus"*
*reddunda in ratione vocare et "semina rerum"*
*appellare suemus et haec eadem usurpare*
*"corpora prima," quod ex illis sunt omnia primis. (DRN 1.54–57)*

*I shall begin to speak to you on the highest principle*
*of the heavens and gods and lay out the first-beginnings of things,*
*from which Nature fashions, grows, and nourishes all things,*
*or to which the same Nature releases them, split apart.*
*These we are accustomed to call "matter" and "bodies generative of things"*
*in giving our account and to name "seeds of things"*
*and to speak of these same entities as "first bodies,"*
*because all things are from these bodies first.*

49. See the related "poverty" passages at *DRN* 1.832 and 3.260.

50. See the remarks of Patricia Parker, who in an essay (1990: 72–73) that pursues (and presses) the distinction between catachresis and metaphor arrives at the following conclusion: "… [*sc.* the distinction between catachresis and metaphor is] not just a distinction based on the presence or absence of an original proper term but the question of conscious control. What is at stake, finally, is both a psychic hierarchy and a social one. And the stakes in the former are no less than the mastery of language itself, the question of whether its movements control or are controlled by the subject in question."

These lines bear quoting because they juxtapose several different facets of *natura*.[51] Although the lexical item *natura* appears only twice, the passage is nonetheless striking, for *natura* in Lucretius often carries the sense of atoms, or physical reality, *tout court*.[52] This passage thus places into a few lines several terms regularly synonymous with *natura: natura* as atoms, *natura* as process (atomic movement), *Natura* as personified agent. Moreover, on a linguistic level, the catalogue of different names for atoms enacts the predicament of catachrestically describing the (atomic basis of the) natural world, for each term—"matter" (*materies*), "bodies generative of things" (*genitalia corpora rebus*), "seeds of things" (*semina rerum*), and "first bodies" (*corpora prima*)—captures an aspect of atoms (i.e., their material or generative nature) at the expense of ignoring their other attributes. That is, an entity that appears initially (and is presented linguistically) as a single type—*primordia rerum*—quickly evolves into a multiplicity of terms. In other words, on both a thematic and a linguistic level, this presentation of *Natura* seems to bear out the quality of catachresis suggested above, namely that it shuttles between an apparently unitary agent (*Natura*) and the material diversity (atoms, atomic movements, the terms describing atoms) that underlies that unitary agent.

Thus, in suggesting that *Natura* is a catachrestic divinity, we emphasize above all two aspects of the trope catachresis: (1) movement (from a proper to an improper sphere) and (2) a unity that yields to (or reveals) diversity. These two features, it should be noted, are not particularly distinct, for—as we have just observed—it may often be a catachrestic linguistic movement that suggests a greater diversity behind an apparent unity. Discussion now turns to demonstrating how these same catachrestic features that have been located in the figure of *Natura* are at work in three separate, though

51. See Campbell 2003: 81: "*Natura* is a description of the process by which all things come into existence and dissolve again into their component atoms, but she moves in the above passage from simply 'the atoms', and becomes a personified controlling force." (Compare, too, Kennedy 2002: 75–117.)

52. See, e.g., *DRN* 1.430–32: "Moreover, there is nothing that you could name that is distinct from all body and separate from void, which would have been discovered as if a third **nature**." (*praeterea nil est quod possis dicere ab omni/ corpore seiunctum secretumque esse ab inani,/ quod quasi tertia sit numero* ***natura*** *reperta*.) *natura* here signifies, as I understand it, "a type of (material) reality," and it indicates, among other things, that one may designate either atoms or void with *natura*. Somewhat similar is the comment at *DRN* 2.1090–92, where *Natura* is at once personified and identified with material nature and its processes: "Which principles, if you grasp them, well-understood, Nature appears/ constantly free, without arrogant rulers,/ to conduct all things herself, of her own accord, by herself, apart from gods." (*quae bene cognita si teneas, Natura videtur/libera continuo, dominis privata superbis,/ ipsa sua* per se *sponte omnia dis agere expers*.)

interconnected, episodes of Lucretius' anthropology: the origins of laguage, the founding of society, and the rise of *religio* ("traditional religion, superstition"). The examination draws not only upon the exploration of catachresis that has been undertaken in this section but also upon earlier observations about the *unus quisque* as at once generic and unique as well as the natural "equality" created by the boundary stones (*termini*) of nature. It is preceded by a brief consideration of the role of performative language within Lucretius' anthropology.

## (ii) Human Catachreses

### *(a) Performative Contexts*

After an introduction (1–54) that praises Epicurus' accomplishments and sets them alongside those of Ceres, Liber, and Hercules, Lucretius' fifth book quickly settles into a lengthy exposition (91–415) about the mortality of the world. At the outset of this demonstration, the poet sets himself within certain known performative contexts. His use of the term *effabor* ("I shall proclaim"), in questioning how his "new, wondrous idea" (97: *res nova miraque*) [i.e., that of the world's mortality] will strike his audience, conjures up the act of "releasing" (*liberare*) a ritual space (*templa*), as performed by the augural priests. Lucretius, it seems, even as he speaks of the difficulty of his task, attempts to "liberate" the minds (*templa mentis*) of his listeners. Only a few lines later (112) the poet compares himself to a Pythian priestess, who "prophesies oracles" (110–12: *fata ... profatur*). These performative roles (particularly that of the augural priest, who creates *templa*) indicate an awareness of the creative power of language that will be further explored in Lucretius' fifth book.

Moreover, the early lines of Lucretius' fifth book introduce a further theme relevant not only to the later development of the book but also to the role and efficacy of performative language described within it. In discussing *Natura* as a catachrestic divinity, we have already pointed to the significance of the atoms and atomic motion that, at times, are identified with *natura* herself but also, whether or not they comprise the whole of "nature," always hover in the background of any usage of *natura*. Lucretius, both in his demonstration of the world's mortality and in the subsequent development of the world, is at pains to dismiss any notion that the world is divine (116) or was created by divine agency for humankind (e.g., 156–59). Instead, he insists, the world is the product of "many atoms" (187: ... *multa ... primordia rerum*) variously driven about over an infinite period (188: *ex infinito ... tempore*).

This dismissal of divine causation and the accompanying introduction of atoms (as a tool to explain the development of natural phenomena), while a standard and well-studied move within Epicurean thinking, bears stressing in the present context because it seems to lie behind several of the "catachrestic" motifs that are traced over the course of this section. In the first instance, the two particular features that have been isolated in the analysis of *Natura* as a catachrestic divinity, movement (from proper to improper) and unity that yields to (or reveals) diversity, are, in many ways, atomic features. Both are aspects of atoms *moving*, coming together to form unities, and then disbanding and moving into other compounds. Moreover, the dismissal of the gods from any direct causal explanation of the physical world is actually instrumental in creating the possibility of performative language. Speaking of the Old Testament God, J. Hillis Miller makes this point well:

> God's *fiat lux* might seem to be the very type or archetype of a proper performative, a way of doing things with words: "And God said, Let there be light: and there was light" (Genesis 1:4). But this by no means fits the definition of an efficacious performative as defined by Austin.... For one thing, it is the report of a putative speech act, not the speech act itself....
>
> Most of all, however, God's *fiat lux* does not fit the definition of a performative because in it knowledge and power, the power of the word, are in perfect congruence, whereas a performative as defined by Austin ... is a form of language in which knowledge and act are asymmetrical, never in perfect harmony. God is omniscient and omnipotent. In him knowledge and power go hand in hand.... A true performative, on the other hand, is a contingent act in the human and social world that makes something happen all right, *though it can never be known for sure beforehand exactly what that something will be* [emphasis added].[53]

While there are clear differences between any Greek or Roman creation story and the divine, biblical "performative" *fiat lux* (which, as Miller stresses, is not a true performative), it is easy to see an "asymmetry" between knowledge and power at work in Lucretius' text. Such asymmetry may be read, for example, in the performative *effabor* ("I shall proclaim") that we have already seen, where

53. Miller 1995: 156–57.

the fact that the verb is in a future tense already suggests the contingency of the utterance (that is, the utterance itself does not *necessarily*, regardless of context, accomplish the act it names).[54] The fact that a performative utterance does not always accomplish its intended purpose—that is, that language may act catachrestically—will loom larger as we move into further examination of the origins of language.

### *(b) Catachrestic Politics*

Turning now to the heart of Lucretius' anthropology, we aim to identify a "catachrestic" structural pattern that repeats itself on several levels—linguistic, political, and religious—in Lucretius' narrative.[55] As Lucretius' account of the origins of the world gets its start by overthrowing gods and replacing them with atoms, so it seems that several of Lucretius' other accounts within the fifth book proceed by presenting (often, as with divine creation, only while eschewing) a governing (monarchical) principle and then replacing that principle with a seemingly more chaotic, individual, democratic principle.[56] These accounts also sow seeds for their own corruption—the misuse of language, the failure to obey just laws, and the re-birth of traditional religion. Figure 4.1 gives a tabular form to the structural argument advanced in the present section. The significance of this pattern, addressed more fully in the conclusion to the present chapter, is, in the first instance, that it provides further weight to the suggestion that the same principles may govern both physical and sociopolitical interaction within Lucretius' text.

54. Compare the statement at Isid. *Orig.* 8.11.90: *fatum autem dicunt esse quidquid dii fantur, quidquid Iuppiter fatur*. ("They say, moreover, that *fatum* is whatever the gods say [*fantur*], whatever Jupiter says [*fatur*].") This remark constitutes a culturally more proximate version of the biblical *fiat lux*.

55. There are many parallels to the scenes studied here in other ancient sources. As the present argument is closely tied to the role of speech acts within Lucretius and Epicureanism, I draw only on sources within the ambit of Epicureanism, or that may plausibly placed within Lucretius' cultural context. Cole 1990, originally published in 1967, presents a (still controversial) argument that many of the motifs in the Epicurean tradition may be traced back to Democritus, a conclusion that must remain speculative due, in large part, to the fragmentary state of our Democritean evidence. Whether or not we accept this argument, though, Cole's book collects a large amount of useful evidence. For further parallels, cf. Lovejoy and Boas (eds.) 1935 and Campbell 2003: 330–53.

56. That is, in the terms presented previously there is a *movement* that changes from a unitary, monarchical principle to a democratic one. The pattern traced here has an obvious connection to Lucretius' anti-teleology, but it is a more fully elaborated pattern than the label "anti-teleological" would suggest.

**Figure 4.1 Chart of Catachrestic Movements**

| Episode | Origins of Language (*ll.* 1028–90) | Founding of Society (*ll.* 1105–60) | Rise of *religio* (*ll.* 1161–1240) |
|---|---|---|---|
| Monarchical Principle Denied | No single name-giver (*ll.* 1041–43) | Overthrow of kings (*l.* 1136) | Attributing causation to the will of the gods (*l.* 1187, 1236–40) |
| Individual Principle (in Response) | Language located in individual speakers (*quisque*) (*l.* 1033) | Spontaneous submission to laws, each individually (*ll.* 1147–49) | *Pleonachos tropos* (*ll.* 526–33); Performative effect of Epicureanism (6.58–67) |
| Difficulties of/ for Individual Response | Language contributes to undermining the senses (*l.* 1133) | Individual stained by fear of punishment (*ll.* 1151–52) | Possible recurrence of universal fear (cf. *ll.* 1218–25, 1236–40) |

Our demonstration of the repeated presence of this catachrestic structure begins with the origins of language, where Lucretius presents himself as emphatically opposed to the idea of a first name-giver:

*Proinde putare* ***aliquem***[57] *tum nomina distribuisse*
*rebus et inde homines didicisse vocabula prima,*
*desiperest. nam cur hic posset cuncta notare*
*vocibus et varios sonitus emittere linguae,*
*tempore eodem* ***alii facere id non quisse putentur****? (DRN 5.1041–45)*

*Accordingly, to think that* ***anyone*** *at that time assigned names*
*to things and from there taught men their first words*
*is madness.*[58] *Why would this man have been able to mark everything*

57. Although Lucretius' *aliquem* is quite vague, the Epicurean Diogenes of Oinanda (fr. 12.III.4–IV.10) launches a similar attack against a first name-giver he explicitly identifies as Hermes.

58. The intended philosophical opponent here has often been debated, but his identity does not much impact the present discussion. Most common is the suggestion that Lucretius is responding to the νομοθέτης ("name-giver, law-maker") of Plato's *Cratylus* (388e4–389a3). See Atherton 2005: 108n20 (for the debate) and the collection of references in Dornseiff

*with sounds and send forth various sounds from his tongue,*
*while at the same time* ***others are thought to have been unable to do this?***[59]

Many analyses of this passage stress the epistemological grounds on which Lucretius denies the possibility of a first name-giver, "whence was implanted a *prolepsis* ["concept"] of its utility" (1047–48: ... *unde insita notities est/ utilitatis* ...)? That argument turns on, as Gordon Campbell puts it, the "somewhat circular" argument that "for language to have developed it was necessary for people to be already communicating with language"; that is, as a *prolepsis* (*notities*) is empirically derived, its existence presupposes the previous existence of its object (here the use of language).[60] But, without denying the importance of such epistemological claims, we may note how emphatically the passage resists the unequal distribution of language ability. Indeed, the lines that immediately follow paint the relationship between the hypothetical

1916: 9–12 (for more on first name-givers). Cicero offers a roughly contemporary account of a first name-giver, dating from perhaps twenty-five years earlier:

> ... fuit quoddam tempus, cum in agris homines passim bestiarum modo vagabantur et sibi victu fero vitam propagabant nec ratione animi quicquam, sed pleraque viribus corporis administrabant ... quo tempore quidam magnus videlicet vir et sapiens cognovit, quae materia esset et quanta ad maximas res opportunitas in animis inesset hominum, si quis eam posset elicere et praecipiendo meliorem reddere; qui dispersos homines in agros et in tectis silvestribus abditos ratione quadam conpulit unum in locum et congregavit et eos in unam quamque rem inducens utilem atque honestam primo propter insolentiam reclamantes, deinde propter rationem atque orationem studiosius audientes ex feris at inmanibus mites reddidit et mansuetos. (Cicero [*De inventione*] 1.2)
>
> There was a certain time when humans used to wander about here and there in fields like beasts and prolong their lives with wild nutriment. They did not conduct anything with the calculation of their minds, but they managed much with the strength of their bodies ... At that moment a certain man, clearly great and wise, recognized what potential there was, how great was the chance for the greatest things in the minds of humanity, if anyone were able to elicit it and make it better by teaching. This man drove together and collected humans, who were scattered in the fields and hidden, in a certain fashion, in forest-homes. He introduced them to every single thing useful and honorable, with them complaining out of unfamiliarity at first; then with them listening more zealously because of their reason and speech (*rationem atque orationem*), he rendered them gentle and tame instead of wild and savage.

59. Campbell 2003: 323 (*ad* 5.1091–1101): "... in the Epicurean scheme ... all people are Promethean, and no art or technology needs an outstanding intelligence as its originator. Further, the Epicurean approach is gradualistic, and so precludes any kind of νομοθέτης, human or divine: all arts developed by the interaction of experience and necessity over time."

60. Campbell 2003: 307.

first name-giver and other early language-learners as one of hostile, monarchical domination:

*__cogere__ item __pluris unus__ victosque __domare__*
*non poterat ut perdiscere nomina vellent (DRN 5.1050–51)*

*Likewise, a __single man__ would be unable __to compel, subdue__*
*and __conquer many__, so that they might desire to learn names/words by heart.*

Early language-learners here appear as wild beasts, being broken in by the first name-giver.[61] The scene draws (and dismisses) a clear portrait of a monarchical structure: one rules over many. And while such a dismissal does not in itself imply the acceptance of an alternative, democratic structure, the ability to produce language is, as we shall see in other passages, spread democratically across individuals.

Against this monarchical foil, Lucretius develops a democratic (that is, equally distributed) potential for language, one that also has a certain catachrestic ability to wander. The "primal scene" for initial language use presents a portrait peopled by several "catachrestic" agents, *Natura* ("Nature"), *utilitas* ("Use/Utility"), *infantia* ("Speechlessness"):[62]

*At varios linguae sonitus Natura subegit*
*mittere, et utilitas expressit nomina rerum,*
*non alia longe ratione atque ipsa videtur*
*protrahere ad gestum pueros infantia linguae,*
*cum facit ut digito quae sint praesentia monstrent.*
*sentit enim vis __quisque suas__ quoad possit __abuti.__ (DRN 5.1028–33)*

*And nature compelled the diverse sounds of the tongue*
*to go forth, and utility coined names for things.*

---

61. Although it is not of central concern to the present argument, the juxtaposition of humans and animals that occurs throughout Lucretius' presentation of the origins of language could be analyzed as providing support for a democratic reading. Early passages within the fifth book tend to stress the distinction between humans and other animals, generally suggesting the superiority of man over other animals. (Cf. esp. 861–62, where some animals [e.g., dogs] are preserved only through their usefulness to humans.) The presentation of the origins of language, by contrast, tends to stress the similarity between animals and humans (e.g., the hypothetical scene within the present passage), although it does maintain some clear distinction (e.g., 1089). That is, the origins of language seem to suggest a more, if not fully, egalitarian ordering of species.

62. There does seem to be some kind of linguistic activity immediately before the fuller development described in the present passage. See *DRN* 5. 1022–23: "... when they **stutteringly**

*For a reason not far different, speechlessness itself*
*seems to draw boys forth to gesture,*
*when it brings about that they indicate what is present with their fingers;*
*for* ***each*** *feels to what extent he can* ***use his own*** *powers.*

The *quisque* ( ... *suus*) ["each [ ... his own]"] in this passage conjures up the *unus quisque*, whom we have discussed elsewhere in this chapter, the generic yet individual figure central to Lucretius' presentation of natural "compacts" (*foedera*) and "boundaries" (*termini*).[63] This figure here receives linguistic potential, a linguistic potential that operates like a seed, with words growing forth from each at the prodding of *natura*.[64] And, not unlike with seeds, while (linguistic) potential may be democratically distributed, (linguistic) development is not uniform.

The linguistic ability that each individual here perceives (*sentit*) appears under the name of the term *abuti* ("to make full use of [*OLD*² s.v. 2]; to put to a wrong use [4]; to use catachrestically [4b]").[65] This term, as these three parenthetical definitions indicate, may be translated variously, but it always seems to conjure up a sense of "use" that transcends a given boundary. Thus,

---

**indicate with voice and gesture** that it is fair for all to pity the weak." (... ***vocibus et gestu cum balbe significarent*** / *imbecillorum esse aequum misererier omnis*).

63. Cf. 5.990: *unus* ... *quisque* ..., which occurs only a few lines before the origins of language.

64. The image of the seed is hardly innocent, as the phrase *semina rerum* is a frequent Lucretian locution for atoms. The notion of the seed helps capture the creative power of atoms. For instances of *semina rerum*, see *DRN* 2.755, 833, 1059, 1072; 6.789, 1093. See, too, Sedley 1998: 193–98, who asserts that, at least by comparison with Epicurus, this turn of phrase represents a Lucretian innovation. Epicurus, although he uses σπέρμα ("seed"), never uses it in the sense of "atom."

65. Commentators (e.g., Ernout-Robin, Bailey) often cite the following passage of Cicero's *De natura deorum* as evidence for the meaning of the term *abuti* ("use to the full"). While this usage bears a clear similarity to that of Lucretius, the Stoic context in which it appears gives it a distinct coloring. That is, the ability here to "abuse" dogs and elephants is part of the natural (divine) order in the world, whereas in Lucretius this "order" is rather the product of atomic interaction:

> Efficimus etiam domitu nostro quadripedum vectiones, quorum celeritas atque vis nobis ipsis adfert vim et celeritatem. nos onera quibusdam bestiis nos iuga imponimus; nos elephantorum acutissimis sensibus nos sagacitate canum ad utilitatem nostram **abutimur**; nos e terrae cavernis ferrum elicimus rem ad colendos agros necessariam, nos aeris argenti auri venas penitus abditas invenimus et ad usum aptas et ad ornatum decoras. (Cicero *ND* 2.60 [151])
>
> In the act of taming them, we cause our horses to draw us, thereby conveying their speed and power to ourselves. We impose burdens to certain beasts, yokes to others; we **borrow** the sharpest senses of elephants, the wisdom of dogs for our own benefit; we draw iron from the bowels of the earth, a commodity necessary for cultivating fields, we uncover deep-hidden veins of bronze, silver, gold, veins both fitted to use and appropriate for decoration.

"using to the full" seems to transgress the boundaries of moderation, while "using wrongly" or "using catachrestically," as explored to a certain extent previously, seem to move beyond the proper. Here this transgression of boundaries may be taken in at least two senses: in the first instance, *abuti* may connote a transgression of the previous limits of human nature, a nature previously "fixed" (after the end of the earth's creative phase, cf. *DRN* 5.826–36) without language;[66] secondarily, though, *abuti* may indicate transgression within language itself.[67] Readings of Epicurean linguistic thought often have postulated that Epicurus (and Lucretius following him) believes in a "natural" meaning for words. But increasingly it is being recognized that Epicurean authors also investigated the development of words beyond these "natural" bounds.[68]

---

66. Gordon Campbell (Campbell 2003: 8–9, *passim* on the "fixity of species") takes the view that, although there is potential variation within other species, the *genus humanum* alone *evolves* within the Lucretian framework. In a different—yet illuminating and comparable—context, Agamben 2004: 33–38 highlights the difficulties of searching for the origins of language and, more specifically, pre-linguistic man: in every conceivable instance, he suggests, locating the origins of language seems to necessitate a change in the "object" of study.

67. It is worth comparing here the remarks of Lucretius' contemporary Varro, who develops a theory of *declinatio*, which governs changes in words (through their "use"), primarily declensions and derivations. The passage below strikingly distinguishes between two types of changes (*declinationes*) in words. *declinatio voluntaria* designates a change by a single author; *declinatio naturalis*, by contrast, designates a change "by common consensus." Varro seems to have in mind standard patterns of nominal declension, e.g., *femina, feminae, feminae, feminam*, etc. With respect to Lucretius, this distinction is fascinating, for Lucretius seems to have a similar sense that certain linguistic developments are "natural," while others (e.g., those that lead to false religious belief and fear) are "voluntary." (Scholars have suggested, with good reason, that parts of Varro's eighth book present an Epicurean, or at least an Epicurean-compatible, linguistic theory. See Blank 2005: 137–38, Dam 1930, Taylor 1975.)

> Declinationum genera sunt duo, voluntarium et naturale. voluntarium est, quo ut cuiusque tulit voluntas declinavit. *sic* tres cum emerunt Ephesi singulos servos, nonnunquam alius declinat nomen ab eo qui vendit Artemidorus, atque Artemam appellat, alius a regione quod ibi emit, ab Ion<i>a Iona, alius quod Ephesi Ephesium, *sic* alius ab alia aliqua re, ut visum est.
>
> Contra naturalem declinationem dico, quae non a singulorum oritur voluntate, sed a com<m>uni consensu. (Varro *LL* 8.21–22)
>
> There are two types of *declinatio* ("derivation; declension" but also "swerve"), voluntary and natural. Voluntary is declension as the will of each dictates. Thus, when three men bought indvidual slaves at Ephesus, sometimes one derives his name from the seller, Artemidorus, and calls him Artemas; another derives the name from the region (because he bought him there), e.g. Ion from Ionia; another calls the slave Ephesius because he was purchased at Ephesus. So one names his slave from one thing, another from another, as seemed best to each.
>
> On the other hand, I call natural *declinatio* ("derivation") that does not arise from the will of individuals but from common agreement.

68. David Sedley's discussion (in Sedley 1973) of the metaphorical extension of "void" (κενός) constitutes one of the first studies of the ways in which Epicurean thought explicitly grappled

One moment where Lucretius explicitly considers the capacity of language to develop catachrestically appears in his narrative of the origins of society, an episode that is considered in greater detail shortly. For now, though, it is enough to note that Lucretius indicates that language can distort and misrepresent information that should be clear from our senses:

*proinde sine incassum defessi sanguine sudent,*
*angustum per iter luctantes ambitionis,*
*quandoquidem sapiunt* ***alieno ex ore*** *petuntque*
*res* ***ex auditis*** *potius quam sensibus ipsis,*
*nec magis id nunc est neque erit mox quam fuit ante. (DRN 5.1131–35)*

*Accordingly, allow that they sweat blood in vain, fatigued,*
*struggling through a narrow path for public office,*
*since they taste/know from* ***a mouth not their own*** *and seek*
*matters more* ***from hearsay*** *than their own senses,*
*but hearsay exists no more now, nor will it more exist soon, than it*
*was before.*

Here it is possible to question the easy division of hearsay (presumably information perceived by our ears) from other sensory information, but for the present purpose it is enough to note that Lucretius apparently believes that we can receive through (the audition of) language a "misrepresentation" of the "natural" information we should receive. In other words, it is possible for language to transcend bounds and function catachrestically.

This rapid trip through the origins of language has suggested that Lucretius replaces a monarchical principle (a first name-giver) with a more chaotic, democratic principle, a seed-like development of language through individuals. It has suggested further that this principle permits a certain "catachrestic" corruption of language, as demonstrated by Lucretius' negative portrayal of hearsay within the rise of society. We turn now to that rise of society to indicate that it follows a pattern not unlike the one just outlined.

Within the larger trajectory of the fifth book, the rise of society occurs after the human race has acquired both language and fire. The primitive society initially shows itself as a meritocracy, where the most intelligent rise to the top, quickly becoming kings:

*Inque dies magis hi victum vitamque priorem*
*commutare novis monstrabant rebus et igni*

***ingenio qui praestabant et corde vigebant.***
*condere coeperunt urbis arcemque locare*
*praesidium* ***reges ipsi*** *sibi perfugiumque,*
*et pecus atque agros divisere atque dedere*
*pro facie cuiusque et viribus ingenioque.*
*posterius res inventast aurumque repertum,*
*quod facile et validis et pulchris dempsit honorem;*
*divitioris enim sectam plerumque sequuntur*
*quamlubet et fortes et pulchro corpore creti. (DRN 5.1105–16)*

*And, with time, those* ***who stood out, flourishing with minds and intelligence,***
*were showing how to change previous food and life*
*for new things and fire. They began,* ***kings themselves,*** *to found cities*
*and to place citadels, as a defense and refuge for themselves,*
*and they divided cattle and fields, granting*
*in accordance with the appearance, might, and intelligence of each.*
*Later, possesions were discovered; gold was found,*
*which easily subtracted honor from both the strong and beautiful,*
*for both those however strong and those however beautiful*
*often follow the set of weathier men.*

Lucretius' meritocratic monarchs, who quickly yield—if not in title, at least in prestige—to plutocrats, show their colors in dividing possessions not equally but in accordance with preferred attributes—appearance, might, and intelligence (1111: *facies, vires, ingenium*)—a practice that no doubt grants most to the monarchs themselves. Yet these monarchs, for all their beauty and intelligence, are unable to hold sway once money and gold are discovered. The chaos inspired by the arrival of gold causes men to seek greater power (and money, cf. *opulenti*) for themselves, apparently out of desire for stability:

*at claros homines voluerunt se atque potentes,*
*ut fundamento stabili fortuna maneret*
*et placidam possent opulenti degere vitam—*
*nequiquam, quoniam ad summum succedere honorem*
*certantes iter infestum fecere viai,*
*et tamen e summo, quasi fulmen, deicit ictos*
*invidia interdum contemptim in Tartara taetra,*
*invidia quoniam, ceu fulmine, summa vaporant*

*plerumque et quae sunt aliis magis edita cumque;*
*ut satius multo iam sit parere quietum*
*quam regere imperio res velle et regna tenere.*
*proinde sine incassum defessi sanguine sudent,*
*angustum per iter luctantes ambitionis,*
*quandoquidem sapiunt alieno ex ore petuntque*
*res ex auditis potius quam sensibus ipsis,*
*nec magis id nunc est neque erit mox quam fuit ante. (DRN 5.1120–35)*

*But they wanted to be famous men, and powerful ones,*
*so that their fortune might abide with a firm basis*
*and they, wealthy, might be able to pass their lives peacefully—*
*in vain, since in competing to reach the highest office*
*they make the path unsafe,*
*and nevertheless from the heights, like a lightning bolt, jealousy*
*at times casts them down scornfully, stricken, into foul Tartarus,*
*since the heights and whatever things are more pronounced than others*
*often are set ablaze with jealousy, like lightning,*
*so that it is much better to obey quietly*
*than to desire to rule things with supreme power and hold kingdoms.*
*Accordingly, allow that they sweat blood in vain, fatigued,*
*struggling through a narrow path for public office,*
*since they know/taste from a mouth not their own and seek*
*matters more from hearsay than their own senses,*
*but hearsay exists no more now, nor will it more exist soon, than it was before.*

These lines describe the competition for political office inspired, at least in part, by a lust for money and jealousy (*invidia*). The insertion of jealousy, twice characterized as functioning "like lightning" (*quasi fulmen*), is striking, for lightning represents a certain conundrum for Lucretius. The connection between lightning and jealousy is, as Bailey remarks, almost proverbial. Yet the proverb, as a passage from Ovid suggests (*Rem. Am.* 369–70: "Jealousy seeks the heights … /lightning bolts sent from Jove's right seek the heights." [*summa petit livor … /summa petunt dextra fulmina missa Iovis*]), has a divine agent at its root. This comparison is unlikely to be a casual slip on the part of the poet, for at *DRN* 6.379–422 he proclaims vehemently that lightning is not a divine phenomenon, suggesting instead (6.6.246–7) that it is produced from clouds. In other words, the comparison of *invidia* to lightning seems to suggest, within a Lucretian framework, a suddenness but also an almost mechanical, physical explanation of human psychology, one that tends to undermine (like the comparison of humans to atoms) a clear, volitional choice to seek political office.[69] Whatever the case, the lightning-bolt of jealousy serves as a Lucretian segue into the destruction of monarchs:

*Ergo* ***regibus occisis*** *subversa iacebat*
*pristina maiestas soliorum et sceptra superba,*
*et capitis summi praeclarum insigne cruentum*
*sub pedibus vulgi magnum lugebat honorem;*
*nam cupide conculcatur nimis ante metutum.*
*res itaque ad summam faecem turbasque redibat,*
*imperium sibi cum ac summatum quisque petebat.*
*inde magistratum partim docuere creare*
*iuraque constituere, ut vellent legibus uti.*
*nam genus humanum, defessum vi colere aevom,*
*ex inimicitiis languebat; quo magis ipsum*
***sponte sua*** *cecidit sub leges artaque iura.*
*acrius ex ira quod enim se* ***quisque*** *parabat*
*ulcisci quam nunc concessumst* ***legibus aequis,***
*hanc ob rem est homines pertaesum vi colere aevom.*
*(DRN 5.1136–60)*

*Therefore,* ***kings were killed;*** *the former majesty*
*of thrones and haughty scepters lay overturned;*
*and the shining crown of the ruler, bloodied,*
*mourned its great office beneath the feet of the common crowd;*
*for lustfully is trampled what previously was feared too much.*

---

with language extending beyond initial "natural" meanings. Recent evidence from *PHerc* 403 fr. 5 col. i 8–14 Armstrong and Fish (= fr. 6.8–14 Sbordone) seems to augment his considerations by providing an Epicurean account of the origins of figural language. See Holmes 2005: 566–68, Mackey 2003.

69. Buried in these lines is also a sentiment ("it is much better to obey quietly than to desire to rule things with supreme power and hold kingdoms") that many take as characteristic of Epicurean politics. Typically, these lines are understood as more or less equivalent to λάθε βιώσας [Usener 551 (in Usener 1887): "Live unnoticed."]. While on any interpretation Epicureanism contains a strain of political quietism, this chapter aims to indicate that Epicurean physical principles have significant political import. Moreover, what Lucretius says here is not "Don't rule! Don't partake in politics!" but rather "Don't *desire* to rule!" At least in this particular formulation, we may read this dictum first and foremost as an injunction to limit our desires. Roskam 2007 examines the reception history of the λάθε βιώσας doctrine.

*So things were returned to the utmost dregs and confusion,*
*when each sought the heights of supreme power for himself.*
*Whence, in part, they taught to create magistracies*
*and to establish a legal system, that they might desire to*
*follow laws.*
*For, the human race, exhausted from leading its live in violence,*
*was wilting from its hostilities; for which reason it*
*all the more fell* ***spontaneously*** *under laws and a confined legal*
*system.*
*For, because* ***each****, out of anger, readied himself to avenge more*
*fiercely than now is permitted by* ***fair legislation****,*
*for this reason it was very tiresome to life live by violence.*

There are many things to be said here, not least that Lucretius, at least in some respects, seems to swerve into recounting Roman history rather than offering a generic, anonymous account of the overthrow of monarchy.[70] But, whatever the historical elements present in this passage, what is most important for the present argument is that, in the first instance, monarchy is overthrown, and second—after a period of anarchy—the *genus humanum*, acting *sua sponte* (like atoms), falls under a set of "equal" laws. In other words, Lucretius' account of the origins of society fits the same general pattern outlined in the origins of language: overthrow of a monarchical principle and replacement of that principle with a democratic one. Moreover, much like the "catachrestic potential" of language to wander and misrepresent our senses, there seems to be a "catachrestic potential" for indivduals under laws to violate those laws, even if that violation carries with it psychological consequences:

*Inde metus maculat poenarum praemia vitae.*
*circumretit enim vis atque iniuria quemque,*
*atque, unde exortast, ad eum plerumque revertit,*
*nec facilest placidam ac pacatam degere vitam*
*qui violat factis communia foedera pacis. (DRN 5.1151–55)*

*Whence fear of punishment stains the prizes of life;*
*for, violence and injustice ensnare each,*

70. See Fowler 1989: 144, who comments that the throne, the scepter, and the crown constitute "the three most important of the regal insignia Rome derived from Etruria" and notes differences between the historical overthrow of kings and Lucretius' account.

*and, from whom the violence arose, to him often it returns,*
*and it is not easy for him to pass a tranquil, peaceful life,*
*who disrupts common compacts of peace with his deeds.*

The rise of society quickly gives way, in Lucretius' account, to the rise of *religio* ("traditional religion; superstition"), which represents a slight deviation from the pattern we have been establishing, in part because *religio* is one of the prime targets of the larger performance of Lucretius' poem. That is, *religio* is not simply a historical practice that has been overthrown, but it is one whose authority Lucretius is in the process of trying to dislodge with his own text. Moreover, the rise of *religio* seems, at least on one interpretation, a product (or a partial product) of the catachrestic potential already located in language. Such, in any case, is the position of one of the most recent commentators on this episode:[71]

> ... religion arises only when reason begins to be applied to nature, and only when language develops, because language is the carrier of the *religio* virus and allows the rapid spread of *religio* by the dissemination of false ideas. This is the argument behind 5.1133–5 where Lucretius claims that politicians behave as they do because they accept the opinions of others rather than applying reason to nature: they are infected by false opinion.[72]

Here Gordon Campbell reads a passage we have already examined—lines that proclaim "hearsay" (*audita*) may lead us away from the truth of our own senses—as indicating that, in the misguided attribution of physical causation to the gods, Lucretius sees language at work no less than in the political context in which these lines appear.

Despite these variations, it is possible to fit Lucretius' account of the rise of *religio* into the general, catachrestic pattern we have outlined to this point. Believing that gods intervene in and conduct the physical world is, according to the language of Lucretius' presentation, submitting to a monarchical principle. We have already seen this fact in a passage (*DRN* 6.54–55) cited earlier,

71. Positing a relationship between *religio* and linguistic error seems to be an Epicurean trope. The phenomenon may be glimpsed not only in Lucretius but also in Philodemus *De pietate* (part I) 519–41, 1176–1217 (Obbink); *De Dis*, Bk. III, 65 fr. 85, col. 2. See Blickman 1989: 174–77.

72. Campbell 2003: 17. The third chapter of Konstan 2008 also emphasizes strongly the ways in which the invention of language erodes trust in direct sensation.

and it is maintained, in a slightly different fashion, in the fuller account of the fifth book:

> *Quippe etenim iam tum divom mortalia saecla*
> ***egregias*** *animo* ***facies*** *vigilante videbant,*
> *et magis in somnis mirando corporis auctu.*
> *his igitur sensum* ***tribuebant*** *propterea quod*
> *membra movere videbantur vocesque* ***superbas***
> *mittere pro* ***facie praeclara*** *et* ***viribus amplis.***
> *aeternamque dabant vitam, quia semper eorum*
> *subpeditabantur facies et forma manebat,*
> *et tamen omnino quod tantis viribus auctos*
> *non temere ulla vi convinci posse putabant.*
> ...
> *Praeterea caeli rationes ordine certo*
> *et varia annorum cernebant tempora verti,*
> *nec poterant quibus id fieret cognoscere causis.*
> *ergo* ***perfugium*** *sibi habebant omnia divis*
> *tradere et* ***illorum nutu*** *facere omnia flecti.*
> ...
> *O genus infelix humanum, talia divis*
> *cum* ***tribuit*** *facta atque iras adiunxit acerbas! (DRN 5.1169–79, 1183–87, 1194–95)*

> *For surely at that time the generations of mortals already*
> *used to see with their waking minds* ***the exceptional faces*** *of gods,*
> *and in sleep they saw them still more with the wondrous bulk of their bodies.*
> *To these, then,* ***they were assigning*** *sensation because*
> *they appeared to move their limbs and to send out their* ***haughty***
> ***voices*** *in accordance with their* ***exceptional appearance*** *and wide strength.*
> *And they were granting them eternal life, because their faces*
> *were (ever) available and their form remained constant,*
> *and finally because they were thinking that those endowed*
> *with such great powers could not easily be conquered by any force.*
> ...
> *Moreover, they were discerning that the rationale of the heavens*
> *with its fixed order and the various times of the years change,*

*and they were unable to come to recognize by what causes this change*
*arose.*
*Therefore, they provided* ***a refuge*** *for themselves: to hand everything*
*over to the gods and to make everything manipulated* ***by their will.***
. . .
*Unhappy race of men, when it* ***assigned*** *such deeds*
*to the gods and added bitter rage!*

The description of the gods here parallels, with uncanny exactness, the description of early kings provided at 5.1105–16. The (images of) gods have exceptional strength and beauty; their voices are described with the adjective *superbus* ("haughty") just as the scepters of kings (5.1137). Just as with the early kings, then, people are inclined to submit to the will (*nutus*)—we might say the (mistaken) monarchical principle—of the gods. Further, this submission to the will of the gods is painted as a performative act (*tribuebant, tribuit*); that is, as in an earlier scene (5.871) where *Natura* is portrayed as the agent, who creatively assigns abilities to the different classes of animals, so here the *genus humanum* creatively assigns power to their (mistaken) visions of gods. Importantly, though, this performative act—the direct language of which is suppressed in the text—is conducted with what J. Hillis Miller calls an "asymmetry" between knowledge and power; the *genus humanum* knows not (at least in the Lucretian presentation) what it does, but it acts nonetheless.

This performative assent to a monarchical principle (i.e., rule of the gods), as indicated earlier, differs from the other "catachrestic" episodes within Lucretius' anthropology in that it represents not only an (imagined) historical phenomenon but also one that Lucretius attempts to combat actively with the present writing of his text. It is thus no surprise that his tale of the rise of *religio* shifts from an account of origins to a present-time account of the wonder and fear inspired by gazing up at the heavens (5.1204–25). This moment represents a clear departure from the other "purely historical" accounts (origins of language, origins of society), but it does not prevent us from suggesting the principle that Lucretius would like to put in place of the "monarchical" account of the gods, for we have several instances both in Lucretius and elsewhere of the so-called *pleonachos tropos*, the multiple explanations of physical phenomena. Consider the following passage:

*Nam quid in hoc mundo sit eorum ponere certum*
*difficile est; sed quid possit fiatque per omne*
*in variis mundis varia ratione creatis,*
*id doceo, plurisque sequor disponere causas,*

*motibus astrorum quae possint esse per omne;*
*e quibus una tamen siet hic quoque causa necessest*
*quae vigeat motum signis; sed quae sit earum*
*praecipere haudquaquamst pedetemptim progredientis.*
*(DRN 5.526–33)*

*For, which of these causes is true in this world,*
*it is difficult to say; but what is possible and comes to be in the sum*
*of all the various worlds created by various principles,*
*this I teach, and I continue to lay out several causes,*
*which can be for the movements of the stars throughout everything;*
*among which causes, one must nevertheless be the cause here,*
*which gives force to the motion of the stars; but it hardly belongs*
*to one proceeding step-by-step to know in advance which of these is true.*

This passage, which may be supplemented by several similar ones, both in Lucretius and other Epicurean texts, presents the "democratic" principle that Lucretius would like to put in place of the "monarchical" explanation on which all things are attributed to the will of the gods. The passage is striking not only because it suggests a more "democratic" view of explanation (that is, one in which several explanations are possible) but also because it suggests a more democratic, partial access to knowledge.[73] The principle of *pleonachos tropos* is, of course, not fully democratic, as it does not permit *any* explanation at all, but instead it requires an explanation (one that pointedly excludes divine agents), which operates within the laws of Epicurean physics.

At this point, it is perhaps unnecessary to add that *religio*, which, at least on one understanding, is founded upon a catachrestic use of language, also sows the seeds for a certain "abuse"—in this instance, the birth and re-birth of religious misconception—as we have seen both with language and society. But Lucretius does discuss this "catachrestic" possibility. As he indicates at *DRN* 6.58, even

73. For further instances of the *pleonachos tropos*, see *DRN* 6.703–11; Epicurus *Ep. ad Hdt.* 79–80, *Ep. ad Pyth.* 86–87; Diogenes of Oenoanda fr. 13.II.12-III.13 (Smith). See, too, Mackey 2006 (which uses the *pleonachos tropos* to explain divergent opinions about atomist theology), Hardie 2009b (which traces the tradition of multiple explanations in later Latin poetry), Hankinson 2013 (which provides a full philosophical account of the *pleonachos tropos*), and Shearin 2014 (which uses the *pleonachos tropos* to unlock the logic of a Lucretian passage on the composition of the soul).

those who have learned well what role the gods play (and what role they do not play) in the administration of the cosmos are nonetheless driven back into "traditional religion" (6.62) through their questioning of the heavens. In other words, it is possible that the very mechanism (5.1211: *rationis egestas* ["want of an account"]) that inspired the first "fall" into religion may strike again.

## *III. Conclusions and Connections (One and Many)*

The preceding section pursued a certain catachrestic pattern—movement from a monarchical principle to a more democratic principle that may be "catachrestically" corrupted—in three episodes within Lucretius' anthropology, with the aim of connecting several strands of the chapter and the larger work more generally. The catachrestic pattern traced in Lucretius' anthropology is striking, at least in part, because it provides a distinctive response to the question of what politics, if any, Lucretius' poem (and Epicureanism more generally) espouses, even if only on an implicit level. On the one hand, traditional accounts of atomism (as well as much of the evidence presented in the present study) tend to emphasize the equality of atoms before natural law; on the other, traditional accounts of the Epicurean school tend to emphasize its pursuit of quietism (a quiet life in the Epicurean Garden) through abstaining from political office.[74] The portrait offered by the catachrestic pattern within Lucretius' anthropology conforms to neither of these positions entirely, although it reflects aspects of both.

There are obvious "democratic" aspects to the catachrestic pattern studied here: inscribing linguistic potential in *each* individual and having individuals fall spontaneously under "equal laws" (*aequae leges*) suggest an egalitarian and nacesent democratic structure. Moreover, the evidence for *termini* reviewed earlier in the chapter infuses a strong sense of equality into any understanding of Lucretius' natural laws. Yet the catachrestic development of language and society seems to indicate a democracy that is hardly permanent; and the Lucretian social and political world, no less than its physical world, is marked by constant movement and change. A review of Lucretius' history of the development of society hints that a number of political configurations may occur, but that none has ultimate stability.

---

74. Cicero's friend, T. Pomponius Atticus, is portrayed, in some ways, as a stereotypical Epicurean in his biography by Cornelius Nepos. At *Atticus* 6, for example, Nepos notes that Atticus "did not seek office ..." (*honores non petiit* ...). On Epicureanism in *Atticus* more generally, see Shearin 2012.

Likewise, there are aspects of Lucretius' presentation that tend to suggest a variety of political quietism. The lines at *DRN* 5.1129–330, "it is much better to obey quietly/ than to desire to rule things with supreme power and hold kingdoms" (... *satius multo iam sit parere quietum/ quam regere imperio res velle et regna tenere*), do present a clear statement against any *desire* for monarchical (or similar) rule, but they are hardly as strong as the famous "live unnoticed" attributed to Epicurus himself. In addition, the evidence presented throughout this chapter concerning the *unus quisque*, the generic yet particular individual who first receives, among other things, linguistic ability, suggests that Lucretius conceived of the human individual, in many, if not all, contexts, as *already part of a group*. The *unus quisque*, for all that Lucretius does seem to believe in individual memories and experiences (*repetentia nostri*), is nonetheless by definition a stereotypical member of the *genus humanum*.

This line of analysis complements the conclusions of the preceding chapter, which stressed the anonymity of individuals as expressed through their personal names. People tend not to appear in the *De rerum natura* apart from groups or classes of individuals, an observation that bears some resemblence to a critique Seneca once lodged against the Epicurean school:

> non sumus sub rege: sibi quisque se vindicat. apud istos quidquid Hermarchus dixit, quidquid Metrodorus, ad unum refertur; omnia quae quisquam in illo contubernio locutus est unius ductu et auspiciis dicta sunt. non possumus, inquam, licet temptemus, educere aliquid ex tanta rerum aequalium multitudine:
>
> pauperis est numerare pecus.
>
> quocumque miseris oculum, id tibi occurret quod eminere posset nisi inter paria legeretur. (Seneca *Ep.* 33.4)
>
> We are not under the rule of a king: each defends himself for himself. Among them <*sc.* Epicureans>, whatever Hermarchus said, whatever Metrodorus said, is referred to a single man; all that anyone in that camp said was spoken under the leadership and auspices of one man. We cannot, I say, though we try, retrieve anything from such a great mass of equal things:
>
> it is a pauper's task to count sheep.
>
> Wherever you cast your eye, this will strike you: what could stand out, if the choice were not among equal things.[75]

---

75. For further discussion of this passage, see Clay 1998: 56, 65 (A reprint of Clay 1983b).

If the lines of investigation pursued within this study have not been misguided, much in what Seneca says is correct. The comparative equality of different Epicureans is both atomic and striking. Yet on at least one point the conclusions of this final chapter suggest a revision to Seneca's statement: his comment captures the anonymity of atoms but not their creative and catachrestic potential. That is, as has been developed in this concluding chapter, in both language and society, Lucretius seems to have as much interest in the way things change and evolve over time as in the fixed qualities of objects. To turn this comment to Seneca's evaluation of the Epicurean school, we may perhaps suggest that the name Epicurus, often itself occluded in the *DRN*, is not so much that of a king but, like the *Natura* that so often appears in Lucretius' text, that of a catachrestic divinity. His name is applied to a broad range of activities and phenomena that, over time, may bear no resemblance to the original Epicurus.

Whatever the validity of this supposition, the catachrestic potential inscribed in language within Lucretius' anthropology may help to clarify the larger significance of speech acts within his poem. As was suggested in the first chapter of this study, speech acts, words that do things, should be of interest to any poet who aims to act upon the mind therapeutically; and Lucretius' anthropology shows perhaps the poet's clearest awareness of the ability of language to misrepresent and confuse reality or, more positively, to cleanse the mind of fear and anxiety. This therapeutic possibility is one of the more hopeful (and most political) outcomes of "doing things with words," and—although there are other possible stopping points—this is where I would like to conclude the present study. Judith Butler, for one, has commented on the ability of the speech act to create "an unanticipated political future":

> ... the speech act, as a rite of institution, is one whose contexts are never fully determined in advance, and ... the possibility for the speech act to take on a non-ordinary meaning, to function in contexts where it has not belonged, is precisely the political promise of the performative ..., one that offers an unanticipated political future. ...[76]

We may suggest something similar for Lucretius' constant borrowing of religious and political language for his rendering of the natural world—that all these speech acts offer the possibility, if hardly the guarantee, of a better future. In the first chapter of this study, it was noted that Epicurus refused to

76. Butler 1997: 161. For a more recent discussion of Butler's ideas on the political implications of performative language, see Butler and Athanasiou 2013.

assent to logical disjunctions in the future tense. As Cicero has it in his *On Fate,*

> necesse est enim in rebus contrariis duabus (contraria autem hoc loco ea dico, quorum alterum ait quid, alterum negat)—ex iis igitur necesse est invito Epicuro alterum verum esse, alterum falsum, ut 'sauciabitur Philocteta' omnibus ante saeculis verum fuit, "non sauciabitur" falsum ... (*De Fato* 37)
>
> It is necessary in the case of two contrary matters (here I call "contrary" these things for which the one affirms something, the other denies it)—of contraries, then, it is necessary, though Epicurus is unwilling to admit it, that one is true, the other false, as (e.g.) "sauciabitur Philocteta" ("Philoctetes will be wounded") was true in all previous ages, "non sauciabitur" ("Philoctetes will not be wounded") was false ...

Perhaps Epicurus—and here we may see Lucretius following him—was unwilling to proclaim either "*sauciabitur Philocteta*" ("Philoctetes will be wounded") or "*non sauciabitur*" ("Philoctetes will not be wounded") because he knew a better future was possible, not just for Philoctetes but for all of us.

# *Works Cited*


Adams, J. N. 1982. *The Latin Sexual Vocabulary*. Baltimore.

Agamben, G. 2011. *The Sacrament of Language: An Archaeology of the Oath*. Stanford. (Originally published in 2008 as *Il sacramento del linguaggio. Archeologia del giuramento.*)

Agamben, G. 2004. *The Open: Man and Animal.* Translated by Kevin Attell. Stanford. (Originally published in 2002 as *L'aperto: l'uomo e l'animale.*)

Agamben, G. 1998. *Homo Sacer: Sovereign Power and Bare Life.* Translated by Daniel Heller-Roazen. Stanford. (Originally published in 1995 as *Homo sacer: Il potere sovrano e la vida nuda.*)

Ahearn, L. M. 2001. "Language and Agency." *Annual Review of Anthropology* 30: 109–37.

Ahl, F. 1985. *Metaformations: Soundplay and Wordplay in Ovid and Other Classical Poets*. Ithaca.

Alberti, A. 1995. "The Epicurean Theory of Law and Justice." In Laks and Schofield (eds.) 1995: 160–91.

Alberti, A. 1990. "Paura della morte e identità personale nell'epicureismo." In A. Alberti (ed.): 151–206. Florence.

Alberti, A. (ed.) 1990. *Logica, mente e persona: studi sulla filosofia antica*. Florence.

Algra, K., J. Barnes, et al. (eds.) 1999. *The Cambridge History of Hellenistic Philosophy*. Cambridge.

Algra, K. et al. (eds.) 1997. *Lucretius and His Intellectual Background*. Amsterdam.

Anderson, W. S. 1960. "Discontinuity in Lucretian Symbolism." *Transactions of the American Philological Association* 91: 1–29.

Annas, J. 1992. *Hellenistic Philosophy of Mind*. Berkeley.

Appel, G. 1909. *De Romanorum precationibus*. Giessen.

Armstrong, D. 1995. "The Impossibility of Metathesis: Philodemus and Lucretius on Form and Content in Poetry." In Obbink (ed.) 1995: 210–32.

Arrighetti, G. (ed.) 1973. *Epicuro: Opere*. 2nd Edition. Turin.

Arweiler, A. 2010. "What Is a Literary Speech Act? Quintilian, John Searle and the Notion of 'Constatives.'" In Fuhrer and Nelis (eds.) 2010: 199–242.

Asmis, E. 2008. "Lucretius' New World Order: Making a Pact with Nature." *Classical Quarterly* 58.1: 141–57.

Asmis, E. 1999. "Epicurean Epistemology." In Algra, Barnes, et al. (eds.) 1999: 260–94.

Asmis, E. 1984. *Epicurus' Scientific Method*. Ithaca.

Asmis, E. 1983. "Rhetoric and Reason in Lucretius." *American Journal of Philology* 104: 36–66.

Asmis, E. 1982. "Lucretius' Venus and Stoic Zeus." *Hermes* 110: 458–70.

Atherton, C. 2009. "Epicurean Philosophy of Language." In Warren (ed.) 2009: 197–215.

Atherton, C. 2005. "Lucretius on What Language Is Not." In Frede and Inwood 2005: 101–36.

Austin, J. L. 1979. *Philosophical Papers*. Third Edition. Oxford.

Austin, J. L. 1975. *How to Do Things with Words*. Second Edition. Cambridge, Mass. (First Edition published 1962.)

Austin, J. L. 1962a. *Sense and Sensibilia*. Oxford.

Austin, J. L. 1962b. "Performatif-Constatif." In *La Philosophie Analytique*, Paris, 1962: 271–304.

Austin, R. G. (ed.) 1960. *Cicero Pro M. Caelio Oratio*. Third Edition. Oxford.

Ax, W. (ed.) 1938. *M. Tullius Cicero. De divinatione, De fato, Timaeus*. Leipzig.

Ax, W. (ed.) 1933. *M. Tullius Cicero. De natura deorum*. Leipzig.

Bailey, C. (ed. and trans.) 1947. *Lucretius: De rerum natura*. 3 vols. Oxford.

Barchiesi, A. and W. Scheidel. (eds.) 2010. *The Oxford Handbook of Roman Studies*. Oxford.

Barnes, J. and M. Griffin (eds.) 1989. *Philosophia Togata*. Oxford. (Corrected reprint, newly entitled *Philosophia Togata I*: 1997.)

Barnes, J. 1996. "Epicurus: Meaning and Thinking." In Giannantoni and Gigante (eds.) 1996: 197–220.

Basanoff, V. 1947. *Evocatio*. Paris.

Beard, M., J. North, and S. Price. 1998. *Religions of Rome*. 2 vols. Cambridge. (= BNP)

Beardsley, M. 1970. *The Possibility of Criticism*. Detroit.

Beer, B. 2009. *Lucrez und Philodem. Poetische Argumentation und poetologischer Diskurs*. Basel.

Begemann, E. 2012. *Schicksal als Argument: Ciceros Rede vom* fatum *in der späten Republik*. Stuttgart.

Bender, J. and D. E. Wellbery (eds.) 1990. *The Ends of Rhetoric: History, Theory, Practice*. Stanford.

Bendlin, A. 2002. "Gemainschaft, Öffentlichkeit und Identität: Forschungsgeschichtl iche Anmerkungen zu den Mustern sozialer Ordnung in Rom." In Egelhaaf-Gaiser and Schäfer (eds.) 2002: 9–40.

Bennett, J. 2010. *Vibrant Matter: A Political Ecology of Things*. Durham, N.C. (USA); London.

Benveniste, É. 1969. *Le vocabulaire des institutions indo-européenes*. 2 vols. Paris.

Benveniste, É. 1966. *Problèmes de linguistique générale I*. Paris.

Bernays, J. 1847. "De emendatione Lucreti." *Rheinisches Museum* 5: 533–87.

Bettini, M. 2012. *Vertere*. Turin.

Bettini, M. 2008. "Weighty Words, Suspect Speech: Fari in Roman Culture." *Arethusa* 41: 313–75.

Blank, D. 2005. "Varro's Anti-Analogist." In Frede and Inwood (eds.) 2005: 210–38.

Blickman, D. R. 1989. "Lucretius, Epicurus, and Prehistory." *Harvard Studies in Classical Philology* 92: 157–91.

Bloom, H. et al. 1979. *Deconstruction and Criticism*. New York.

Braund, S. H. (ed.) 1996. *Juvenal: Satires, Book 1*. Cambridge.

Braund, S. H. and C. Gill (eds.) 1997. *The Passions in Roman Thought and Literature*. Cambridge.

Brown, R. D. 1987. *Lucretius on Love and Sex*. Leiden.

Bruns, I. 1884. *Lucrezstudien*. Freiburg.

Brunschwig, J. 1995. *Études sur les philosophies hellénistiques: Épicurisme, stoïcisme, scepticisme*. Paris.

Brunschwig, J. (ed.) 1978. *Les Stoïciens et leur logique*. Paris.

Brunschwig, J. 1977. "Epicure et le problème du 'langage privé.'" *Revue des sciences humaines* 43: 157–77. (Reprinted in Brunschwig 1995: 43–68.)

Brunt, P. A. 1988a. *The Fall of the Roman Republic and Related Essays*. Oxford.

Brunt, P. A. 1988b. "Amicitia in the Late Republic." Revised version. In Brunt 1988a: 351–81. (Originally published in 1965 in *Proceedings of the Cambridge Philological Society* 11: 1–20.)

Burnyeat, M. F. 1978. "The Upside-Down Back-to-Front Sceptic of Lucretius IV 472." *Philologus* 122.2: 197–206.

Butler, J. 1997. *Excitable Speech: A Politics of the Performative*. London.

Butler, J. 1993. *Bodies that Matter: On the Discursive Limits of* "Sex." London.

Butler, J. and A. Athanasiou. 2013. *Dispossession: The Performative in the Political*. Malden, Mass.: Cambridge.

Butterfield, D. 2013. *The Early Textual History of Lucretius'* De rerum natura. Cambridge.

Cabisius, G. 1985. "Social Metaphor and the Atomic Cycle in Lucretius." *Classical Journal* 80.2: 109–20.

Campbell, G. 2003. *Lucretius on Creation and Evolution. A Commentary on* De rerum natura 5.772–1104. Oxford.

Canfora, L. 1993. *Vita di Lucrezio*. Palermo.

Carroll, N. 1992. "Art, Intention, and Conversation." In Iseminger (ed.) 1992: 97–131.

Cavell, S. 1995. *Philosophical Passages: Wittgenstein, Emerson, Austin, Derrida*. Oxford.

Christensen, J. 2010. "First-Person Futures in Homer." *American Journal of Philology* 131: 543–71.

Classen, C. J. (ed.) 1986. *Probleme der Lukrezforschung*. Hildesheim.

Clay, D. 1998. *Paradosis and Survival: Three Chapters in the History of Epicurean Philosophy*. Ann Arbor.

Clay, D. 1996. "An Anatomy of Lucretian Metaphor." In Giannantoni and Gigante (eds.) 1996: 2.779–93. (Reprinted as Chapter 9 in Clay 1998.)

Clay, D. 1986. "The Cults of Epicurus." *Cronache Ercolanesi* 16: 12–28. (Reprinted as Chapter 5 in Clay 1998.)

Clay, D. 1983a. *Lucretius and Epicurus*. Ithaca.

Clay, D. 1983b. "Individual and Community in the First Generation of the Epicurean School." In *ΣΥΖΗΤΗΣΙΣ* 1983: 1.255–79. (Reprinted as Chapter 4 in Clay 1998.)

Clay, D. 1969. "De Rerum Natura: Greek Physis and Epicurean Physiologia (Lucretius 1.1–148)." *Transactions of the American Philological Association* 100: 31–47. (Reprinted as Chapter 7 in Clay 1998.)

Cole, T. 1990. *Democritus and the Sources of Greek Anthropology*. Atlanta. (Originally published in 1967.)

Colin, J. 1955. "Sobriquets de femmes dans la Rome alexandrine (d'après Cicéron et Lucrèce)." *Revue Belge de Philologie et d'Histoire* 33: 853–76.

Conte, G. B. 1994a. *Genres and Readers*. Baltimore.

Conte, G. B. 1994b. "Instructions for a Sublime Reader: Form of the Text and Form of the Addressee in Lucretius's De rerum natura." In Conte 1994a: 1–34. (Originally published in Italian in 1990.)

Cooper, J. 2012. *Pursuits of Wisdom: Six Ways of Life in Ancient Philosophy from Socrates to Plotinus*. Princeton.

Corbeill, A. 1996. *Controlling Laughter: Political Humor in the Late Roman Republic*. Princeton.

Culler, J. 2007. *The Literary in Theory*. Stanford.

Culler, J. 2000. "Philosophy and Literature: The Fortunes of the Performative." *Poetics Today* 21: 503–19. (Republished in Culler 2007: 137–65)

Culler, J. 1983. *On Deconstruction: Theory and Criticism after Structuralism*. Ithaca.

Culler, J. 1981. *The Pursuit of Signs: Semiotics, Literature, Deconstruction*. Ithaca.

Dam, R. J. 1930. De Analogia observationes in Varronem grammaticamque Romanorum. Dissertation, Amsterdam.

Davies, H. S. 1931–32. "Notes on Lucretius." *The Criterion* 11: 25–42. (Reprinted in Classen (ed.) 1986: 273–90.)

De Lacy, P. 1969. "Limit and Variation in the Epicurean Philosophy." *Phoenix* 23.1: 104–13.

De Lacy, P. 1957. "Process and Value: An Epicurean Dilemma." *Transactions of the American Philological Association* 88: 114–26.

De Lacy, P. 1939. "The Epicurean Analysis of Language." *American Journal of Philology* 60: 85–92.

De Lacy, P. and E. De Lacy (eds.) 1978. *Philodemus: On Methods of Inference*. Naples.

de Man, P. 1979. *Allegories of Reading: Figural Language in Rousseau, Nietzsche, Rilke, and Proust*. New Haven.

Depew, M. and D. Obbink (eds.) 2000. *Matrices of Genre: Authors, Canons, and Society*. Cambridge, Mass.

Derrida, J. 2005. *Sovereignties in Question: The Poetics of Paul Celan*. New York.
Derrida, J. 1988. *Limited Inc*. Edited by G. Graff. Evanston, Ill.
Derrida, J. 1984. "My Chances/Mes Chances: A Rendezvous with Some Epicurean Stereophonies." In Smith and Kerrigan (eds.) 1984: 1–32.
Derrida, J. 1979. "Living On: Border Lines." In Bloom et al. 1979: 75–176.
Dickey, E. 2002. *Latin Forms of Address from Plautus to Apuleius*. Oxford.
Diels, H. (ed.) 1923. *T. Lucreti Cari De rerum natura*. Berlin. 2 vols.
Dionigi, I. 1988. *Lucrezio. Le parole e le cose*. Bologna.
Dorandi, T. (ed.) 2013. *Diogenes Laertius: Lives of Eminent Philosophers*. Cambridge.
Dornseiff, F. 1916. *Buchstabenmystik*. Leipzig.
Dryden, J. 1995. *The Works of John Dryden*. Ware (Hertfordshire). (Date of original publication varies: Preface to Sylvae dates from 1685.)
Dudley, D. R. (ed.) 1965. *Lucretius*. London.
Dupont, F. 1992. *Daily Life in Ancient Rome*. Translated by Christopher Woodall. (Originally published in 1989 as *La vie quotidienne du citoyen romain sous la République*.)
Dyck, A. R. (ed.) 2003. *Cicero. De natura deorum: Liber I*. Cambridge.
Edmunds, L. 2001. *Intertextuality and the Reading of Roman Poetry*. Baltimore.
Egelhaaf-Gaiser, U. and A. Schäfer (eds.) 2002. *Religiöse Vereine in der römischen Antike*. Tübingen.
Eigler, U., U. Gotter, et al. (eds.) 2003. *Formen römischer Geschichtsschreibung von den Anfängen bis Livius. Gattungen, Autoren, Kontexte*. Stuttgart.
Everson, S. 1994a. "Epicurus on Mind and Language." In Everson (ed.) 1994b: 74–108.
Everson, S. 1994b. *Language: Companions to Ancient Thought 3*. Cambridge.
Erler, M. 2009. "Epikureismus als Orakelphilosophie: Orakel und mantik in der hellenistischen Philosophie." In Witte and Diehl (eds.) 2009: 53–66.
Erler, M. 1994. "Lucrez." In Flashar (ed.) 1994: 381–490.
Erler, M. 1993. "Philologia Medicans. Wie die Epikureer die Texte ihres Meisters lasen." In Kullmann and Althoff (eds.) 1993: 281–303.
Ernout, A. (ed.) 1978. *Lucrèce, de la nature*. Paris. (Revised by C. Rambaux.)
Ernout, A. and L. Robin. 1926. *Lucrèce* De Rerum Natura: *Commentaire Exégétique et Critique*. Paris. 3 vols.
Fallot, J. 1977. *Il piacere e la morte nella filosofia di Epicuro*. Turin.
Farrell, J. 2001. *Latin Language and Latin Culture from Ancient to Modern Times*. Cambridge.
Farrington, B. 1965. *Science and Politics in the Ancient World*. New York. (Originally published 1939.)
Feeney, D. 1998. *Literature and Religion at Rome: Cultures, Contexts, and Beliefs*. Cambridge.
Felman, S. 2002. *The Scandal of the Speaking Body: Don Juan with J. L. Austin, or Seduction in Two Languages*. Translated by Catherine Porter. Stanford. (Originally published in 1980 as *Le scandale du corps parlant*.)

Fenves, P. 2001. *Arresting Language: From Leibniz to Benjamin*. Stanford.

Fish, J. and K. R. Sanders (eds.) 2011. *Epicurus and the Epicurean Tradition*. Cambridge.

Fish, S. 1982. "With the Compliments of the Author: Reflections on Austin and Derrida." *Critical Inquiry* 8: 693–721.

Fitzgerald, W. 1995. *Catullan Provocations: Lyric Poetry and the Drama of Position*. Berkeley.

Flashar, H. (ed.) 1994. *Die Philosophie der Antike*, Band 4: Die Hellenistische Philosophie. Basel.

Fleischmann, W. B. 1960. "Alexander Pope and Lucretian 'Anonymity.'" *Neophilologus* 44.3: 216–17.

Fletcher, R. 2012. "Epicurus's Mistresses: Pleasure, Authority, and Gender in the Reception of the Kuriai Doxai in the Second Sophistic." In Holmes and Shearin (eds.) 2012: 52–88.

Fordyce, C. J. (ed.) 1961. *Catullus. A Commentary*. Oxford.

Foucault, M. 1984. *Histoire de la sexualité III: Le souci de soi.* Paris.

Fowler, D. 2002. *Lucretius on Atomic Motion: A Commentary on* De rerum natura *2.1–332*. Oxford.

Fowler, D. 2000. "The Didactic Plot." In Depew and Obbink (eds.) 2000: 205–19.

Fowler, D. 1997. "Epicurean Anger." In Braund and Gill (eds.) 1997: 16–35.

Fowler, D. 1995. "From Epos to Cosmos: Lucretius, Ovid, and the Poetics of Segmentation." In Innes et al (eds.) 1995: 3–18.

Fowler, D. 1989. "Lucretius and Politics." In Barnes and Griffin (eds.) 1989: 120–50.

Fowler, D. 1986. Review of Asmis 1984. *Journal of Hellenic Studies* 106: 227–31.

Fraenkel, E. 1916. "Zur Geschichte des Wortes fides." *Rheinisches Museum* 71: 187–99.

Frede, D. and B. Inwood (eds.) 2005. *Language and Learning: Philosophy of Language in the Hellenistic Age*. Proceedings of the Ninth Symposium Hellenisticum. Cambridge.

Frede, M. 1978. "Principles of Stoic Grammar." In Rist (ed.) 1978: 27–75.

Fredrick, D. (ed.) 2002. *The Roman Gaze: Vision, Power, and the Body*. Baltimore.

Freyburger, G. 1986. *Fides: Étude sémantique et religieuse depuis les origines jusqu'à l'époque augustéenne*. Paris.

Friedländer, P. 1941. "Pattern of Sound and Atomistic Theory in Lucretius." *American Journal of Philology* 62: 16–34.

Fuhrer, T. and D. Nelis. (eds.) 2010. *Acting with Words: Communication, Rhetorical Performance and Performative Acts in Latin Literature*. Heidelberg.

Gaisser, J. H. (ed.) 2007. *Catullus*. Oxford. (Oxford Readings in Classical Studies)

Gale, M. (ed.) 2007. *Lucretius*. Oxford. (Oxford Readings in Classical Studies)

Gale, M. (ed.) 2004a. *Latin Epic and Didactic Poetry: Genre, Tradition and Individuality*. Swansea.

Gale, M. 2004b. "The Story of Us: A Narratological Analysis of Lucretius' De rerum natura." In Gale (ed.) 2004a: 49–71.

Gale, M. 2001a. *Lucretius and the Didactic Epic*. London.

Gale, M. 1994. *Myth and Poetry in Lucretius*. Cambridge.

Gallivan, P. 1992. "The Nomenclature Patterns of the Roman Upper Class in the Early Empire: A Statistical Analysis." *Antichthon* 26: 51–79.

Garani, M. 2007. *Empedocles* Redivivus: *Poetry and Analogy in Lucretius*. New York.

Gaulin, M. 2007. Review of Koch 2005. Bryn Mawr Classical Review 2007.08.17. (Available online at http://ccat.sas.upenn.edu/bmcr/2007/2007–08-17.html; link accessed 14 March 2013.)

Giannantoni, G. 1990. *Socratis et Socraticorum Reliquiae*. 4 vols. Naples. (SSR)

Giannantoni, G. and M. Gigante. (eds.) 1996. *Epicureismo Greco e Romano: Atti del Congresso Internazionale, Napoli, 19–26 Maggio 1993*. Naples.

Gigandet, A. 1998. *Fama Deum. Lucrèce et les raisons du mythe*. Paris.

Gigandet, A. 1996. "Natura Gubernans (Lucrèce V, 77)." In Lévy (ed.) 1996: 213–25.

Gigante, M. 1975. "Philosophia Medicans in Filodemo." *Cronache Ercolanesi* 5: 53–61.

Gigon, O. (ed.) 1978. *Lucrèce*. Geneva. (Entretiens 24, Fondation Hardt)

Gillespie, S. and P. Hardie (eds.) 2007. *The Cambridge Companion to Lucretius*. Cambridge.

Gladhill, C. W. 2008. *Foedera*: A Study in Roman Poetics and Society.Dissertation, Stanford University.

Glidden, D. 1985. "Epicurean Prolepsis." *Oxford Studies in Ancient Philosophy* 3: 175–217.

Glidden, D. 1983a. "Epicurean Semantics." In *ΣΥΖΗΤΗΣΙΣ* 1983: 1.185–226.

Glidden, D. 1983b. "Skeptic Semiotics." *Phronesis* 28: 213–55.

Glidden, D. 1981. "Sensus and Sense Perception in the De rerum natura." *California Studies in Classical Antiquity* 12: 155–81.

Glidden, D. 1979. "Epicurus on Self-Perception." *American Philosophical Quarterly* 16: 297–306.

Goldberg, J. 2009. *The Seeds of Things: Theorizing Sexuality and Materiality in Renaissance Representations*. New York.

Goldschmidt, V. 1978. "Remarques sur l'origine épicurienne de la 'prénotion.'" In J. Brunschwig (ed.) 1978: 155–69.

Gordon, C. A. 1962. *A Bibliography of Lucretius*. London.

Gordon, P. 2002. "Some Unseen Monster: Rereading Lucretius on Sex." In Fredrick (ed.) 2002: 86–109.

Gotter, U. 2003. "Die Vergangenheit als Kampfplatz der Gegenwart: Catos (konter) revolutionäre Konstruktion des republikanischen Errinerungsraums." In Eigler, Gotter, et al. (eds.) 2003: 115–34.

Govaerts, S. 1986. *Lucrèce* De rerum natura. *Index verborum*. Liège.

Greenblatt, S. 2011. *The Swerve: How the World Became Modern*. New York; London.

Grimal, P. 1978. "Le poème de Lucrèce en son temps." In Gigon (ed.) 1978: 233–70.

Guthrie, W. K. C. 1940. Review of Farrington 1939. *Classical Review* 54: 34–5.

Hadot, P. 1987. *Exercises spirituels et philosophie antique.* Second Edition. Paris. (Originally published 1981; Translated into English as *Philosophy as a Way of Life* in 1995.)

Hadszits, G. D. 1918. "Lucretius as a Student of Roman Religion." *Transactions of the American Philological Association* 49: 145–60.

Hankinson, R. J. 2013. "Lucretius, Epicurus, and the Logic of Multiple Explanations." In Lehoux, Morrison, and Sharrock (eds.) 2013: 69–97.

Hardie, P. 2009a. *Lucretian Receptions: History, the Sublime, Knowledge*. Cambridge.

Hardie, P. 2009b. "Lucretian Multiple Explanations and Their Reception in Latin Didactic and Epic." In Hardie 2009a: 231–63.

Hardie, P. 2002. *Ovid's Poetics of Illusion*. Cambridge.

Hardie, P. 1986. *Virgil's* Aeneid: *Cosmos and Imperium*. Oxford.

Heinze, R. 1929. "Fides." *Hermes* 64: 140–66.

Heinze, R. (ed.) 1897. *T. Lucretius Carus: De rerum natura Buch III*. Leipzig.

Hellegouarc'h, J. 1963. *Le vocabulaire des relations et des partis politiques sous la République*. Paris.

Henderson, J. G. W. 2006. "Oxford Reds": *Classic Commentaries on Latin Classics*. London.

Henderson, J. G. W. 2004. *Morals and Villas in Seneca's* Letters: *Places to Dwell*. Cambridge.

Henderson, J. G. W. 1999. *Writing Down Rome: Satire, Comedy, and Other Offences in Latin Poetry*. Oxford.

Henderson, J. G. W. 1998. *Fighting for Rome: Poets & Caesars, History & Civil War*. Cambridge.

Henderson, J. G. W. 1997. *Figuring Out Roman Nobility: Juvenal's Eighth* Satire. Exeter.

Hersch, K. 2010. *The Roman Wedding: Ritual and Meaning in Antiquity*. Cambridge.

Hexter, R. and D. Selden (eds.) 1992. *Innovations of Antiquity*. New York.

Heyworth, S. J. (ed.) 2007. *Classical Constructions: Papers in Memory of Don Fowler, Classicist and Epicurean*. Oxford. (With P. G. Fowler and S. J. Harrison.)

Holford-Strevens, L. 2002. "Horror vacui in Lucretian Biography." *Leeds International Classical Studies 1.1*: 1–23. (Available online at http://lics.leeds.ac.uk/2002/200201.pdf; link accessed 14 March 2013.)

Holmes, B. 2005a. "Daedala Lingua: Crafted Speech in De Rerum Natura." *American Journal of Philology* 126: 527–85.

Holmes, B. 2005b. Review of Campbell 2003. *Aestimatio* 2: 142–62.

Holmes, B. and W. H. Shearin (eds.) 2012. *Dynamic Reading. Studies in the Reception of Epicureanism*. Oxford.

Humphries, R. (trans.) 1968. *Lucretius, The Way Things Are: The De Rerum Natura of Titus Lucretius Carus*. Bloomington.

Hutchinson, G. O. 2001. "The Date of the De rerum natura." *Classical Quarterly* 51.1: 150–62.

Innes, D. et al. (eds.) 1995. *Ethics and Rhetoric: Classical Essays for Donald Russell on His Seventy-Fifth Birthday*. Oxford.

Iseminger, G. (ed.) 1992. *Intention and Interpretation*. Philadelphia.

Jobst, F. 1907. *Über das Verhältnis zwischen Lukretius und Empedokles*. Munich.

Johnson, B. 1980. *The Critical Difference: Essays in the Contemporary Rhetoric of Reading*. Baltimore.

Johnson, W. R. 2000. *Lucretius and the Modern World*. London.

Kajanto, I. 1965. *The Latin Cognomina*. Helsinki.

Kajanto, I. 1977. "On the Chronology of the Cognomen in the Republican Period." In *L'Onomastique latine*, Paris, 1977: 63–70.

Kaser, M. 1971. *Das römische Privatrecht.* Erster Abschnitt: Das altrömische, das vorklassische und klassische Recht. München.

Kaster, R. A. 1988. *Guardians of Language: The Grammarian and Society in Late Antiquity*. Berkeley.

Keen, R. 1985. "Lucretius and His Reader." *Apeiron* 19: 1–10.

Kennedy, D. F. 2013a. *Antiquity and the Meanings of Time: A Philosophy of Ancient and Modern Literature*. London.

Kennedy, D. F. 2013b. "The Political Epistemology of Infinity." In Lehoux, Morrison, and Sharrock (eds.) 2013: 51–67.

Kennedy, D. F. 2002. *Rethinking Reality: Lucretius and the Textualization of Nature*. Ann Arbor.

Kennedy, D. F. 2000a. "Making a Text of the Universe: Perspectives on Discursive Order in the De Rerum Natura of Lucretius." In Sharrock and Morales (eds.) 2000: 205–25.

Kennedy, D. F. 2000b. "Bluff Your Way in Didactic: Ovid's Ars Amatoria and Remedia Amoris." *Arethusa* 33: 159–76.

Kennedy, D. F. 1993. *The Arts of Love: Five Studies in the Discourse of Roman Love Elegy*. Cambridge.

Kennedy, D. F. 1992. "'Augustan' and 'Anti-Augustan': Reflections on Terms of Reference." In Powell (ed.) 1992: 26–58.

Kenney, E. J. (ed.) 1984. *Lucretius: De rerum natura. Book III*. Corrected Reprint. Cambridge.

Kenney, E. J. 1977. *Lucretius*. Oxford. (*Greece and Rome* New Surveys in the Classics)

Kenney, E. J. 1970. "Doctus Lucretius." *Mnemosyne* 4.23: 366–92.

Klingner, F. 1952. "Philosophie und Dichtkunst am Ende des zweiten Buches Lucrez." *Hermes* 80: 3–31.

Knox, P. and C. Foss (eds.) 1998. *Style and Tradition: Studies in Honor of Wendell Clausen*. Stuttgart; Leipzig.

Koch, R. 2005. *Comment peut-on être dieu? La secte d'Épicure*. Paris.

Konstan, D. 2008. *A Life Worthy of the Gods. The Materialist Psychology of Epicurus*. Las Vegas.

Konstan, D. 1997. *Friendship in the Classical World*. Cambridge.

Kornemann, E. 1901. "Coloniae." *Paulys Realencyclopädie der classischen Altertumswissenschaft* 4.7: 511–88.

Kullmann, W. and J. Althoff (eds.) 1993. *Vermittlung und Tradierung von Wissen in der griechischen Kultur.* Tübingen.

Kumaniecki, K. (ed.) 1969. *M. Tulli Ciceronis De Oratore*. Stuttgart.

Lachmann, K. 1850. *In T. Lucreti Cari De Rerum Natura Libros Commentarius*. Berlin.

Lachmann, K. (ed.) 1850. *T. Lucreti Cari De Rerum Natura Libri Sex*. Berlin.

Laks, A. and M. Schofield. (eds.) 1995. *Justice and Generosity*. Cambridge.

Laursen, S. 1995. "The Early Parts of Epicurus, On Nature, 25th Book." *Cronache Ercolanesi* 25: 5–109.

Laursen, S. 1997. "The Later Parts of Epicurus, On Nature, 25th Book." *Cronache Ercolanesi* 27: 5–82.

Lausberg, H. 1990. *Handbuch der literarischen Rhetorik. Eine Grundlegung der Literaturwissenschaft*. Third Edition. Stuttgart.

Lehoux, D. 2013. "Seeing and Unseeing, Seen and Unseen." In Lehoux, Morrison, and Sharrock (eds.) 2013: 131–51.

Lehoux, D., A. D. Morrison, and A. Sharrock. (eds.) 2013. *Lucretius: Poetry, Philosophy, Science*. Oxford.

Lévy, C. (ed.) 1996. *Le Concept de Nature à Rome: La Physique*. Paris.

Lieberg, G. 1982. *Poeta creator: Studien zu einer Figur der antiken Dichtung*. Amsterdam.

Lieberg, G. 1962. *Puella Divina. Die Gestalt der göttlichen Geliebten bei Catull im Zusammenhang der antiken Dichtung.* Amsterdam.

Liechtenhan, E. (ed.) 1968. *Marcelli De Medicamentis Liber (Post Maximilianum Niedermann Iteratis Curis)*. Berlin. (Corpus Medicorum Latinorum V)

Long, A. A. 2006. *From Epicurus to Epictetus: Studies in Hellenistic and Roman Philosophy*. Oxford.

Long, A. A. 1997. "Lucretius on Nature and the Epicurean Self." In Algra et al. (eds.) 1997: 125–39.

Long, A. A. 1971. "Aisthesis, Prolepsis and Linguistc Theory in Epicurus." *Bulletin of the Institue of Classical Studies* 18: 114–33.

Long, A. A. and D. Sedley. 1987. *The Hellenistic Philosophers*. 2 Vols. Cambridge. (= LS)

Lovejoy, A. O. and G. Boas (eds.) 1935. *Primitivism and Related Ideas in Antiquity*. Baltimore; London. (Reprint: 1997)

Lowrie, M. 2009. *Writing, Performance, and Authority in Augustan Rome*. Oxford.

Loxley, J. 2007. *Performativity*. London.

Lyne, R. 1980. *The Latin Love Poets*. Oxford.

MacKendrick, P. L. 1952. "Roman Colonization." *Phoenix* 6: 139–146.

Mackey, J. L. 2006. "Saving the Appearances: The Phenomenology of Epiphany in Atomist Theology." Available online at Princeton/Stanford Working Papers in Classics (http://www.princeton.edu/~pswpc/papers/authorMZ/mackey/mackey.html; link accessed 14 March 2013).

Mackey, J. L. 2003. "New Evidence for the Epicurean Theory on the Origin of Language." M. St. Thesis, Oxford (Christ Church).

Madvig, J. N. 1832. *De aliquot lacunis codicum Lucretii*. Copenhagen.

Maltby, R. 1991. *A Lexicon of Ancient Latin Etymologies*. Leeds.

Manetti, G. (ed.) 1996. *Knowledge through Signs: Ancient Semiotic Theories and Practices.* Turnhout.

Manuwald, A. 1972. *Die Prolepsislehre Epikurs.* Bonn.

Marcovich, M. 1999. (ed.) *Diogenes Laertius Vitae Philosophorum.* Vol. 1: Libri 1–10. Stuttgart.

Marshall, P. K. 1983. "Servius." In Reynolds (ed.) 1983: 385–88.

Martin, J. 1963. *T. Lucreti Cari De Rerum Natura Libri Sex.* Fifth Edition. Leipzig.

Martin, R. P. 1989. *The Language of Heroes: Speech and Performance in the* Iliad. Ithaca.

Mayer, R. G. 1990. "The Epic of Lucretius." *Papers of the Leeds International Latin Seminar* 6: 35–43.

Meillet, A. 1922. "Latin credo et fides." *Mémoires de la Société de Linguistique de Paris* 22: 215–18.

Mewaldt, J. 1927. "T. Lucretius Carus." *Paulys Realencyclopädie der classischen Altertumswissenschaft* 13.2: 1659–83.

Miller, J. Hillis. 2007. "Performitivity as Performance/Performativity as Speech Act: Derrida's Special Theory of Performativity." *South Atlantic Quarterly* 106: 219–35.

Miller, J. Hillis. 2005. *Literature as Conduct: Speech Acts in Henry James.* New York.

Miller, J. Hillis. 2001. *Speech Acts in Literature.* Stanford.

Miller, J. Hillis. 1995. *Topographies.* Stanford.

Minyard, J. D. 1985. *Lucretius and the Late Republic.* Leiden.

Mitsis, P. 1993. "Committing Philosophy on the Reader: Didactic Coercion and Reader Autonomy in De Rerum Natura." *Materiali e Discussioni per L'Analisi dei Testi Classici* 31: 111–28.

Mommsen, T. 1887. *Römisches Staatsrecht.* 3 vols. Third Edition. Leipzig. (Reprint: Basel, 1952)

Momigliano, A. 1941. Review of Farrington 1965. *Journal of Roman Studies* 31: 149–57.

Monet, A. (ed.) 2003. *Le jardin romain: Epicurisme et poésie à Rome: Mélanges offerts à Mayotte Bollack.* Villeneuve d'Ascq.

Morel, P.-M. 2009. *Épicure: la nature et la raison.* Paris.

Müller, K. 1975. *T. Lucreti Cari De Rerum Natura Libri Sex.* Zürich.

Münzer, F. 1931. "C. Memmius (8)." *Paulys Realencyclopädie der classischen Altertumswissenschaft* 15.1: 610–16.

Mynors, R. A. B. 1958. *C. Valerii Catulli Carmina.* Oxford. (Oxford Classical Text)

Mynors, R. A. B. (ed.) 1969. *P. Vergili Maronis Opera.* Oxford. (Oxford Classical Text)

Nagy, G. 1990. "The Crisis of Performance." In Bender and Wellbery (eds.) 1990: 43–59.

Neubecker, A. J. (ed.) 1986. *Philodemus. Über die Musik IV. Buch.* Naples.

Newman, J. K. 1967. *The Concept of Vates in Augustan Poetry.* Brussels.

Nichols, J. H., Jr. 1976. *Epicurean Political Philosophy: The* De Rerum Natura *of Lucretius.* Ithaca.

Nicolet, C. 1977. "L'Onomastique des groupes dirigeants sous la République." In *L'Onomastique latine*, Paris, 1977: 45–61.

Niedermann, M. and E. Liechtenhan (eds.) 1968. *Marcelli De medicamentis liber*. Translated (into German) by J. Kollesch and D. Nickel. Berlin.

Nietzsche, F. 1988. *Friedrich Nietzsche: Sämtliche Werke. Kritische Studienausgabe in 15 Bänden*. Second Edition. München. (KSA of G. Colli and M. Montinari.)

Norden, E. 1995. *Aus altrömischen Priesterbüchern*. Leipzig. (Originally published 1939.)

Norden E. 1898. *Die antike Kunstprosa vom VI. Jahrhundert v. Chr. bis in die Zeit der Renaissance*. 2 vols. Leipzig.

O'Hara, J. J. 2007. *Inconsistency in Roman Epic. Studies in Catullus, Lucretius, Vergil, Ovid and Lucan*. Cambridge.

O'Hara, J. J. 1996. *True Names: Vergil and the Alexandrian Tradition of Etymological Wordplay*. Ann Arbor.

O'Keefe, T. 2005. *Epicurus on Freedom*. Cambridge.

Oliensis, E. 1998. *Horace and the Rhetoric of Authority*. Cambridge.

Obbink, D. (ed.) 1995. *Philodemus and Poetry: Poetic Theory and Practice in Lucretius, Philodemus, and Horace*. Oxford.

Orelli, J. C. 1827. Review of H. Meyer (ed.) *Ciceronis Orator* (Leipzig, 1827). *Jahrbücher für Philologie und Paedagogik*, ser. 3.4: 84–91.

Parker, A. and E. K. Sedgwick. (eds.) 1995. *Performativity and Performance*. New York.

Parker, P. 1990. "Metaphor and Catachresis." In Bender and Wellbery (eds.) 1990: 60–73.

Passannante, G. 2011. *The Lucretian Renaissance. Philology and the Afterlife of Tradition*. Chicago.

Piccaluga, G. 1974. *Terminus: I segni di confine nella religione romana*. Rome.

Pigeaud, J. 1981. *La maladie de l'âme: étude sur la relation de l'âme et du corps dans la tradition médico-philosophique antique*. Paris.

Plasberg, O. (ed.) 1922. *M. Tullius Cicero. Academicorum reliquiae cum Lucullo*. Stuttgart.

Porter, J. I. 2004. "Vergil's Voids." *Helios* 31: 127–56.

Porter, J. I. 2003. "Lucretius and the Poetics of Void." Available online at Porter's personal website (https://sites.google.com/site/jamesivanporter/articles; link accessed 14 March 2013.) Originally published in Monet (ed.) 2003: 197–226.

Porter, J. I. 1996. "In Search of an Epicurean Aesthetics." In Giannantoni and Gigante (eds.) 1996: 611–28.

Powell, J. G. F. (ed. and trans.) 1990. *Cicero: Laelius, On Friendship & The Dream of Scipio*. Warminster.

Pötscher, W. 1978a. "'Numen' und 'numen Augusti.'" *Aufstieg und Niedergang der römischen Welt* 2.16.1: 355–92.

Pötscher, W. 1978b. "Das römische Fatum—Begriff und Verwendung." *Aufstieg und Niedergang der römischen Welt* 2.16.1: 393–424.

Powell, A. (ed.) 1992. *Roman Poetry and Propaganda in the Age of Augustus*. Bristol.

Purmann, H. 1846. *Quaestionum Lucretianarum specimen*. Bratislava.

Reeve, M. D. 2005. "The Italian Tradition of Lucretius Revisited." *Aevum* 79: 115–64.

Reeve, M. D. 1980. "The Italian Tradition of Lucretius." *Italia medioevale e umanistica* 23: 27–48.

Regenbogen, O. 1932. *Lucrez: seine Gestalt in seinem Gedicht*. Leipzig; Berlin.

Reiley, K. C. 1909. *Studies in the Philosophical Terminology of Lucretius and Cicero.* New York.

Reinhardt, T. 2008. "Epicurus and Lucretius on the Origins of Language." *Classical Quarterly* 58.1: 127–40.

Reinhardt, T. 2005. "The Language of Epicureanism in Cicero: The Case of Atomism." *Proceedings of the British Academy* 129: 151–77.

Reynolds, L. D. (ed.) 1983. *Texts and Transmission: A Survey of the Latin Classics.* Oxford.

Rist, J. (ed.) 1978. *The Stoics*. Berkeley.

Roller, D. W. 1970. "Gaius Memmius: Patron of Lucretius." *Classical Philology* 65: 246–8.

Rorty, R. 1991. *Essays on Heidegger and Others*. Cambridge.

Roskam, G. 2007. "Live Unnoticed" *Λάθε βιώσας: On the Vicissitudes of an Epicurean Doctrine*. Leiden.

Rothacker, E. (ed.) 1962. *Archiv für Begriffsgeschichte*. Vol. 7. Bonn.

Rouse, W. H. D. (ed. and trans.) 1992. *Lucretius: De Rerum Natura*. Cambridge, Mass. (Revised by M. F. Smith.)

Rüpke, J. 1990. *Domi militiae: Die religiöse Konstruktion des Krieges in Rom.* Stuttgart.

Sallmann, K. 1962. "Studien zum philosophischen Naturbegriff der Römer mit besonderer Berücksichtigung des Lukrez." In E. Rothacker (ed.) 1962: 140–284.

Salomies, O. 1987. *Die römischen Vornamen: Studien zur römischen Namengebung.* Helsinki.

Salmon, E. T. 1969. *Roman Colonisation under the Republic*. London.

Salway, B. 1994. "What's in a Name? A Survey of Roman Onomastic Practice from c. 700 B.C. to A.D. 700." *Journal of Roman Studies* 84: 124–45.

Scheid, J. 1989–90. "Hoc anno immolatum non est. Les aléas de la voti sponsio." *Scienze dell'Antichità. Storia Archeologia Antropologia* 3–4: 773–83.

Schiesaro, A. 2007a. "Lucretius and Roman Politics and History." In Gillespie and Hardie (eds.) 2007: 41–58.

Schiesaro, A. 2007b. "Didaxis, Rhetoric and the Law in Lucretius." In Heyworth (ed.) 2007: 63–90.

Schiesaro, A. 1990. *Simulacrum et imago: Gli argomenti analogici nel* De rerum natura. Pisa.

Schrijvers, P. H. 1999. *Lucrèce et les sciences de la vie.* Leiden.

Schrijvers, P. H. 1970. *Horror ac Divina Voluptas. Études sur la poétique et la poésie de Lucrèce*. Amsterdam.

Searle, J. R. 1983. "The World Turned Upside Down." *New York Review of Books*. Vol. 30, no. 16. (October 27, 1983)

Searle, J. R. 1977. "Reiterating the Differences: A Reply to Derrida." *Glyph* 1: 198–208.

Searle, J. R. 1969. *Speech Acts: An Essay in the Philosophy of Language*. Cambridge.

Sedgwick, E. 1993. "Queer Performativity: Henry James's The Art of the Novel." *Gay and Lesbian Quarterly* 1.1: 1–16.

Sedley, D. 2003. "Lucretius and the New Empedocles." *Leeds International Classical Studies* 2.4.

Sedley, D. 1998. *Lucretius and the Transformation of Greek Wisdom*. Cambridge.

Sedley, D. 1996. "Aristotle's De Interpretatione and Ancient Semantics." In Manetti 1996: 87–108.

Sedley, D. 1983. "Epicurus' Refutation of Determinism." In *ΣΥΖΗΤΗΣΙΣ* 1983: 1.11–51.

Sedley, D. 1977. "Diodorus Cronus and Hellenistic Philosophy." *Proceedings of the Cambridge Philological Society* 203: 74–120.

Sedley, D. 1973. "Epicurus, On Nature Book XXVIII." *Cronache Ercolanesi* 3: 5–83.

Seele, A. 1995. *Römische Übersetzer: Nöte, Freiheiten, Absichten*. Darmstadt.

Segal, C. 1990. *Lucretius on Death and Anxiety. Poetry and Philosophy in* De rerum natura. Princeton.

Selden, D. L. 2007. "Ceveat lector: Catullus and the Rhetoric of Performance." In Gaisser (ed.) 2007: 490–559. (Expanded and updated version of Selden 1992.)

Selden, D. L. 1998. "Alibis." *Classical Antiquity* 17: 289–412.

Selden, D. L. 1992. "Ceveat lector: Catullus and the Rhetoric of Performance." In Hexter and Selden (eds.) 1992: 461–512.

Sells, M. A. 1994. *Mystical Languages of Unsaying*. Chicago.

Sgarbi, R. 1992. "A proposito del lessema latino 'Fetiales.'" *Aevum* 66.1: 71–78.

Sharrock, A. and H. Morales (eds.) 2000. *Intratextuality*. Oxford.

Shearin, W. H. Forthcoming. "The Play of the Lucretian Signifier: Saussure's Anagrammatic Readings of De rerum natura." In D. O'Rourke (ed.) *Lucretius in Theory*.

Shearin, W. H. 2014. "Concealed Pleasure: Lucretius De rerum natura 3.237–42." *Classical Quarterly* 64.1: 183–96.

Shearin, W. H. 2012. "Haunting Nepos: Atticus and the Performance of Roman Epicurean Death." In Holmes and Shearin (eds.) 2012: 30–51.

Skutsch, O. (ed.) 1985. *The* Annals *of Q. Ennius*. Oxford.

Smith, J. H. and W. Kerrigan (eds.) 1984. *Taking Chances: Derrida, Psychoanalysis, and Literature*. Baltimore.

Snyder, J. M. 1980. *Puns and Poetry in Lucretius'* De rerum natura. Amsterdam.

Souter, A. 1949. *A Glossary of Later Latin to 600 A.D.* Oxford.

Stanford, W. B. 1936. *Greek Metaphor*. Oxford.

Striker, G. 1974. "Κριτήριον τῆς ἀληθείας." *Nachrichten der Akademie der Wissenschaften in Göttingen*. Philologische-historische Klasse 2: 47–110.

*ΣΥΖΗΤΗΣΙΣ*. 1983. *Studi sull'Epicureismo Greco e Romano Offerti a Marcello Gigante*. 2 Vols. Naples.

Taylor, D. J. 1975. *Declinatio: A Study of the Linguistic Theory of Marcus Terentius Varro*. Amsterdam.

Timpanaro, S. 2005. *The Genesis of Lachmann's Method*. Chicago. (Originally published in Italian in 1981 and reprinted with corrections in 1985.)

Terada, R. 2005. "Seeing Is Reading." In Marc Redfield (ed.) 2005. *Legacies of Paul de Man*. Romantic Circles Praxis Series. (Available online at http://www.rc.umd.edu/praxis/deman/terada/terada.html; link accessed 14 March 2013.)

Townend, G. B. 1978. "The Fading of Memmius." *Classical Quarterly* 28.2: 267–83.

Usener, H. 1887. *Epicurea*. Leipzig. (Reprint with Italian Translation and Notes by Ilaria Ramelli: Milan, 2002.)

Veyne, P. 1988. *Roman Erotic Elegy. Love, Poetry, and the West*. Trans. D. Pellauer. Chicago.

Voelke, A.-J. 1993. *La philosophie comme thérapie de l'âme: études sur la philosophie hellénistique*. Fribourg.

Volk, K. 2002a. *The Poetics of Latin Didactic: Lucretius, Vergil, Ovid, Manilius*. Oxford.

Volk, K. 2002b. Review of Gale 2001a. Bryn Mawr Classical Review 2002.03.42. (Available online at http://ccat.sas.upenn.edu/bmcr/2002/2002–03-42.html; link accessed 14 March 2013.)

Volk, K. 2001. Review of Johnson 2000. Bryn Mawr Classical Review 2001.01.17. (Available online at http://ccat.sas.upenn.edu/bmcr/2001/2001–01-17.html; link accessed 14 March 2013.)

Warren, J. 2006. "Epicureans and the Present Past." *Phronesis* 51.4: 362–87.

Warren, J. (ed.) 2009. *The Cambridge Companion to Epicureanism*. Cambridge.

Waszink, J. H. 1949. "Two Remarks on Roman Poets." *Mnemosyne* 4: 68–71.

West, D. 1991. Review of Dionigi 1988. *Gnomon* 63: 647–9.

West, D. 1982. Review of Snyder 1982. *Classical Review* 32: 25–7.

West, D. 1969. *The Imagery and Poetry of Lucretius*. Edinburgh. (Reprint: Norman (Okla.), 1994.)

Wissowa, G. 1912. *Religion und Kultus der Römer*. Second Edition. Munich.

Wiseman, T. P. 2004. *The Myths of Rome*. Exeter.

Witte, M. and J. F. Diehl (eds.) 2009. *Orakel und Gebete. Interdiziplinäre Studien zur Sprache der Religion in Ägypten, Vorderasien und Griechenland in hellenistischer Zeit*. Tübingen.

Wittgenstein, L. 1997. *Philosophical Investigations*. Second Edition. Translated by G. E. M. Anscombe. (Originally published 1958.)

Wormell, D. E. W. 1960. "Lucretius: The Personality of the Poet." *Greece & Rome* 7: 54–65. (Reprinted in Classen (ed.) 1986: 17–28.)

Zetzel, J. 1998. "De re publica and De rerum natura." In Knox and Foss (eds.) 1998: 230–247.

Zetzel, J. (ed.) 1995. *Cicero* De re publica. *Selections*. Cambridge.

Ziegler, K. 1936. "Der Tod des Lucretius." *Hermes* 61: 421–40.

# Index